ADDITIONS

ADDITIONS

From the Editors of **Fine Homebuilding**®

The Taunton Press

The Taunton Press
Inspiration for hands-on living®

The Taunton Press, Inc., 63 South Main Street, PO Box 5506, Newtown, CT 06470-5506
e-mail: tp@taunton.com

Distributed by Publishers Group West

JACKET AND COVER DESIGN: Ann Marie Manca
INTERIOR DESIGN: Cathy Cassidy
LAYOUT: Cathy Cassidy
COVER PHOTOGRAPHERS: Front cover: Charles Bickford. Back cover: Charles Miller, courtesy *Fine Homebuilding*, © The Taunton Press, Inc. (left); Steve Culpepper, courtesy *Fine Homebuilding*, © The Taunton Press, Inc. (top right); © Mark F. Heffron (center); Scott Gibson, courtesy *Fine Homebuilding*, © The Taunton Press, Inc. (bottom right).

LIBRARY OF CONGRESS CATALOGING-IN-PUBLICATION DATA
Additions : design ideas for great American houses.
       p. cm.
"From the editors of Fine homebuilding."
   ISBN 1-56158-655-2
   1. Dwellings--Remodeling. 2. Buildings--Additions--Design and
construction. I. Fine homebuilding.

TH4816 .A284 2002
690'.8--dc21

                          2002154945

Printed in the United States of America
10 9 8 7 6 5 4 3 2 1

The following manufacturers/names appearing in *Additions* are trademarks: Dumpster™.

Contents

Introduction

At some point in my home's 200-year history, a previous owner added 4 ft. across one end. This addition created a space 4 ft. wide and 30 ft. long, which, if you think about it, is pretty odd. Who adds a hallway to their house?

As if the odd size weren't bad enough, the addition was built on a peanut-brittle foundation—stones laid along the ground with concrete drizzled over them—and framed with old doors, stood up, one next to the other, and nailed together. Given that the frost line here in Connecticut is at 42 in., the foundation heaved, the wall sagged, and for at least 80 years, cold air and small animals enjoyed ready access to the house.

After ignoring the problem for as long as I could, I recently guttted the wall and started digging a real foundation. When I'm done with this project, I will have rebuilt my odd little addition in the same place and the same size as before—sturdier, more conventionally, but much the same. You might wonder why? Well, I'm a fan of houses that grow over time, and I want to preserve some record of this addition.

Like the pencil lines on a doorjamb chronicling a child's height through the years, additions are a visual history of a home's growth. They tell a story about who lived in the house and how their needs, or their resources, changed over time. And sometimes, a series of additions can even manage a kind of fubsy charm that wouldn't have been possible had the whole thing been built at once. That's what I'm hoping for my place. But not all additions fit the telescoping model of my old cape. Some will dwarf the house they accompany; others will blend seamlessly.

As this book attests, additions come in all shapes and sizes. Collected here are 25 articles from past issues of *Fine Homebuilding*. Written by builders and architects from all over the country, these projects offer dozens of useful lessons for adding onto your house, whether all at once or bit by bit.

—*Kevin Ireton*, editor-in-chief

PART ONE

A Coastal Remodel Triumphs over Limits

FIVE YEARS AGO, I GOT A CALL FROM DAMIENNE AND VLAD Zeman of Westport, Connecticut. They had bought a starter home a few years earlier and needed design services for a simple renovation of that house. It was on a nice site in Saugatuck Shores, a coastal New England neighborhood of previously unheated vacation shacks slowly being rehabbed into year-round dwellings on postage-stamp-size lots facing saltwater.

Renovation becomes Reinvention

Initially, their needs were relatively modest. They wanted to raise their home to the code-compliant flood level of 13 ft. above mean high tide, to renovate the home's interior, and to redecorate its exterior. The house's plan formed a modified "T" shape; the main section had bedrooms above and living space below, with a simple one-story wing launching off the side containing the kitchen and garage (see photo above).

A new look amid a regulatory maze. Dormers accent this 12-in-12 pitch roof, which extends over both main wings, creating a saltbox. However, complex local and federal restrictions held the height of the new roof within tight regulatory limits. A nonconforming status as well as budgetary concerns also made it necessary to recycle the two-story portion of the original building (see photo facing page). The footprint is all that remains of the one-story wing. Photos taken at A on floor plan.

A NEW HOUSE MAKES THE BEST OF AN OLD FOOTPRINT

Regulations kept this project from expanding much beyond the original dimensions, so the new plan uses the T–shape of the existing house. The two-story north-south section of the house was recycled with a steeper roof adding a dramatic note to the bedrooms. A master bedroom with a cathedral ceiling tops off the rebuilt east–west section of the house with an upgraded kitchen and garage on the first floor.

SPECS

BEDROOMS: **3**

BATHROOMS: **2½**

SIZE: **2,800 sq. ft.**

COST: **n/a**

COMPLETED: **2000**

LOCATION: **Westport, Connecticut**

ARCHITECT: **Duo Dickinson**

BUILDER: **C & J Construction**

During the design process, the Zemans ultimately decided to address their long-term housing needs. This decision was due partly to their changing needs; they already had one child, and another was on the way by the time construction commenced. The change in plans meant the initial budget grew, and the impact of our expanded design services had to fall within strict local and federal regulations.

Because of our pre-existing, nonconforming status and budgetary concerns, we recycled the two-story portion of the building. The other wing was to be renovated into a more functional kitchen and garage with a new master suite above.

Homeowners Unwind the Red Tape

Virtually every aspect of this home was barely within the building and zoning requirements that would have been imposed on a naked lot. We simply could not have built a new home of this design in this location.

Zoning regulations allowed a small expansion of the existing footprint. Similarly, the height of the building was limited to 26 ft.; the height restriction had to take into account that we were raising the house approximately 3 ft. above its current height to comply with the Coastal Area Management Code.

Custom construction becomes the obvious choice for anyone wanting to build in this context. How do you deal with such extraordinary budgetary and regulatory limits and still create something special? The answer is simple: the investment of time. By investing their own time into working out solutions, homeowners can often solve problems that could be more quickly dealt with by an infusion of raw manpower (and thus dollars).

Second Floor

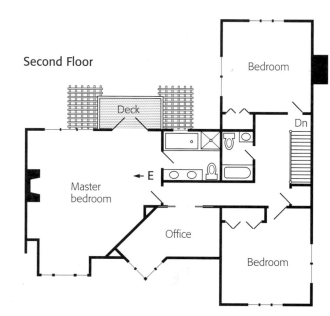

First Floor

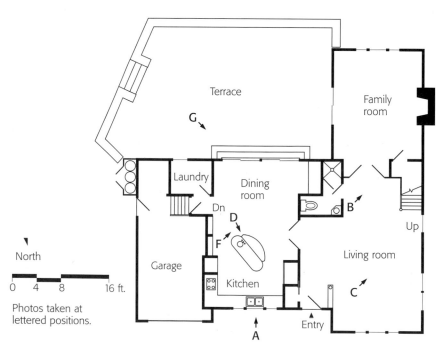

North

0 4 8 16 ft.

Photos taken at lettered positions.

Because our firm was working on an hourly basis, our time was tightly monitored. Fortunately, the Zemans took the bull by the horns, tirelessly massaging zoning and building department issues through a maze of hearings, meetings, correspondence, and informal discussions among engineers, surveyors, town officials, the builder, and me. Their aggressive attention to this part of the process allowed us to obtain building permits and approvals in a relatively short time.

Similarly, they assaulted their strident budgetary limitations in a variety of ways. First, they picked C & J Construction of Madison, Connecticut, an out-of-town builder that offered a better price because they were located outside of Fairfield County's aggressive pricing; second, they directly purchased all the custom millwork, many of the lighting fixtures, and much of the hardware. The homeowners also threw themselves into the process of writing checks to suppliers, being responsible for deliveries, and following up on loose ends.

As many homebuilding veterans know, this approach can be a shortcut to errors and finger-pointing. Homeowners' inexperience in construction management often translates into disappointments and recriminations with architects and builders. In this case, though, the Zemans' unstinting attention to detail and seemingly inexhaustible energy ensured that this project would deliver an excellent bang for the buck.

Recycling the Original Plan Pays Off

The homeowners' enthusiasm alone didn't make this project a success. It also required a design that would take advantage of the building's existing plan. Mindful of building and zoning requirements, we had to work almost exclusively within the house's original "T" formation, which conveniently sep-

arated the formal front of the house from the informal back, with the service areas to one side on the first floor (see floor plans facing page). The structure is now a 2,800-sq.-ft., three-bedroom, 2½-bath house. The upstairs bedrooms stake out the ends of each wing, with baths nestled between them.

On the rear exterior of the house, the intersecting wings were a natural location for a large-scale terrace off the dining room, which looks over the salt marsh (see photo below). At the front of the house, the intersection of the wings became the main entry.

Mindful of building and zoning requirements, we had to work almost exclusively within the house's original "T" formation.

Patio doors frame a lovely view. A built-in cabinet serves the dining area. Patio doors provide access to a large terrace overlooking the salt marsh. Photo taken at F on floor plan.

Round and oval windows are placed in key locations. The oval window at the bottom of the stairs adds visual interest, allows natural light to enter, and provides an outside view. Photo taken at C on floor plan.

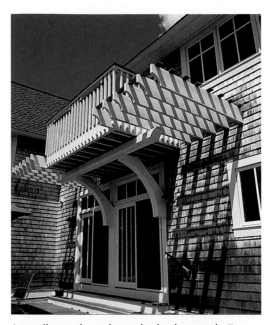

A cantilevered porch overlooks the marsh. Two large brackets and a large beam support a cantilevered porch off the master bedroom. Trellises with contoured tails flank the porch. Photo taken at G on floor plan.

New Roof Animates the Exterior

The most obvious change in the house is the roof. The roofline swoops just above the entry, which has a curved ceiling. And the 6-in-12 pitch roof gave way to a new 12-in-12 pitch, which extends over the entry and the garage bay, creating a saltbox for the two intersecting wings.

Not so obvious are the laminated beams used to pick up some of the point loads generated at the intersections of roofs. The new 12-in-12 pitch roofs needed only engineered collar ties along with prefabricated steel hurricane ties installed at all rafter/plate connections to provide stiffness.

The gable dormer over the garage provides visual interest as well as additional headroom in the master bedroom; another smaller dormer projects at a 45-degree angle with a multifaceted roof just above the entrance. To distinguish these dormers further, the head casings of the individual windows in the gabled dormers rise as the gable roof ascends; the sills on the angled dormer windows fall as they progress down the main roof. To make the dormers more a part of the roof, they were sided with tongue-and-groove red-cedar siding, in contrast with the surrounding house, which is finished in white-cedar shingles.

The same head/sill interplay evident in the dormers is echoed in the heads of the kitchen windows below. In addition, a curved transom window with splayed sides above the entry door complements the swoop in the roof. Round and oval windows are set at key points on the exterior: at gable peaks, at the bottom of the stairs (see photo facing page) and in bathrooms.

In addition, the muntin patterns on many windows follow a cruciform motif, with the vertical and horizontal bars creating square panes at the top and elongated panes below. The round and oval windows are similar, but rather than the cruciform, they have a simple crosshairs pattern.

We added several other elements on the exterior to enrich the overall form. At the entry roof overhang, a large knee brace was added. In the rear, the tails of the cantilevered framing that support the trellis/porch off the master bedroom were carefully sculpted (see photo above) and supported by an octagonal beam and a pair of knee braces.

The tapered chimney uses both trim and a flared, lead-coated copper cap to add distinction to this focal point. The eaves of the house are rendered to complement and unify the rest of the exterior trim details as well as the roof forms.

ABOVE Durable and cost-effective materials throughout. Painted drywall, hardwood floors, and solid-wood trim combine successfully to create a modest yet elegant interior, as in this family-room sitting area. Photo taken at B on floor plan.

FACING PAGE The island defines the kitchen area. With its curved front wrapped in cherry wainscot, the island serves as a divider between the kitchen and dining room. The bar top is teak, and the countertops are granite. Photo taken at D on floor plan.

ABOVE Curved-top armoire complements the angled ceiling. This stylish built-in closet neatly tucks into the voids alongside the fireplace in the master bedroom. Photo taken at E on floor plan.

SOURCES

C & J Construction
Madison, Connecticut
(203) 740-0532

Modest Choices Elsewhere Permit an Elegant Kitchen

The kitchen, whose details the owners took on with gusto, became the heart of an open floor plan (see photo above left). Cherry frame-and-panel cabinetry was used throughout, including the front panels of the refrigerator and dishwasher, making them less conspicuous.

The island is set at an angle for easier access. The curved front is wrapped in cherry wainscot, the bar top is oiled teak, and the countertops are granite.

We integrated the kitchen cabinets into the window trim over the kitchen sink. Two cabinets with glass-panel doors fill the niche created by the smaller windows, which flank the larger window over the kitchen sink.

In addition to the kitchen, the interior finishes of hardwood flooring and solid-wood trim used in the den and throughout the house are durable and simple (see photo above right). Painted drywall became dramatic in the master-bedroom suite after we fleshed out the shapes resulting from the intersecting rooflines (see photo facing page).

Duo Dickinson is an architect in Madison, Connecticut. He is the author of the forthcoming book, *The House You Build* (The Taunton Press, Inc., 2004).

A Town House Opens Up in Philadelphia

AFTER BUYING HIS NEW HOUSE, TOM PEDERSON CAME TO US with two simple requests: He wanted more space and more natural light. Not a terribly tall order, on the face of it, until you consider that the house was a tiny (11 ft. wide, 30 ft. long) two-story row house in central Philadelphia. Our challenge was to create flexible, open living space, to improve circulation in plan and section, and to capture plenty of natural light within the existing building. Because the house is located in a historic district, we needed to accomplish all this without substantially altering the street facade. Our approach was to open the interior space and then to break out of the existing envelope by adding a third floor (see photo p. 14) and a small addition on the back.

A third floor addition and terrace transform an 18th-century row house into a modern light-filled space. Photo taken at A on floor plan.

A third-floor addition gives the house extra room, light and a skyline view. The author designed this third-story addition to step back from street sight lines in compliance with historic-district code and to give the owner a small deck. Photo taken at B on floor plan.

The Problems of Limited Space and Light

Built in 1752, the existing house was wedged into a small, narrow street—the garbage truck's wheels rub the curb on both sides of the street. The dark, cramped interior of the existing house was chopped up by partition walls, which made a small footprint feel still smaller. The available daylight was severely limited as well. In front, houses on the opposite side of the street are only 12 ft. away; to the rear, the backs of buildings on the next block are 18 ft. away. The house's windows admitted only a minimum amount of light. The living room and the dining room/kitchen were on the first floor, with two small bedrooms and a bathroom on the second level. Typical of row houses, the kitchen—at the back of the house—was dark and gloomy.

Designing Compact Additions that Capture More Natural Light

The ability to add to the existing house was restricted both by cost considerations and by Philadelphia zoning ordinances, so we planned to make the most of every square foot. We also envisioned both additions as collectors of much-needed natural light, in addition to providing more usable space. After researching the zoning requirements, we determined the maximum buildable area and height requirements for the site and designed a new third-floor bedroom suite with a terrace (see photos above and facing page) and an 8-ft. by 9-ft. three-story wing on the back side.

On the third floor, we maximized the glass in the north and south walls because no openings were permitted by code in the other two, which are firewalls separating

Large third-floor windows help to light the lower floors. An open stairwell channels sunlight from the top floor to the rest of the house. Sliding doors hidden behind the closet at left provide privacy for the bedroom. Photo taken at C on floor plan.

the row house from those on each side. We installed oversize wood double-hung windows that flood the bedroom suite with natural light and offer views in both directions. We used stock sizes of Marvin Magnum-grade windows (see Sources) in the front because their style was approved by the Philadelphia Historic Commission. We opened the ceiling to the gable roof and separated the roof from the north and south walls with clerestory windows at the eaves. We also designed an operable transom window to maximize light and to enhance air circulation. The light entering

on this level also enables the stair to act as a lightwell, bringing natural light to the center of the house.

To bring much-needed light to the kitchen and dining area, we designed the smaller addition at the back of the house. The addition steps back at the northwest corner, creating a notched-out corner that is mostly glass, with six Kolbe & Kolbe double-hung wood windows joined together vertically from the first floor to the third (see floor plan p. 18).

LEFT A small 9-ft. by 8-ft. addition at the rear of the house creates a brighter kitchen space. A three-story notch above the rear door contains windows that illuminate the cherry cabinets and marble countertops of the new kitchen. Photo taken at F on floor plan.

ABOVE Sliding doors of
sandblasted glass and fir serve
as optional privacy screens.
Traffic from the open stairway and
library beyond can be blocked off
from the second-floor sitting room,
which can double as a guest room.
Photo taken at E on floor plan.

Reinventing the Character of the Existing Space

To start with, we decided that the living and kitchen/dining rooms should remain on the first floor. A new library, a bathroom, and a sitting room that could double as a guest room would occupy the second level, leaving the master suite secluded on the third floor.

As we developed the building plan, we attempted to suggest rather than to define the living spaces. We removed all existing interior partitions—one of the few advantages of a space that's only 11 ft. wide—and limited our use of new partitions as much as possible. The first floor is free of partition walls and open front to back; the kitchen is separated from the dining area by a 36-in. high cabinet with a marble countertop (see photo p. 16). Because the kitchen is open to the living and dining areas, we worked with materials and colors that were warm and

MODEST ADDITIONS AND AN OPEN STAIR IMPROVE THE CIRCULATION OF A ROW HOUSE

To maximize the amount of space, the architects eliminated unnecessary partitions and added an unobtrusive third-floor addition that conformed to historic-district codes. Windows in a small addition at the rear of the house bring much-needed light to the interior.

SPECS

BEDROOMS: 1

BATHROOMS: 2

HEATING SYSTEM: Gas-Fired Hot Air

SIZE: 1,555 sq. ft., including basement

COST: $152.00 per sq. ft.

COMPLETED: 1995

LOCATION: Philadelphia, Pennsylvania

ARCHITECT: John Andrews and David Bae

BUILDER: Hanson General Contracting of Philadelphia

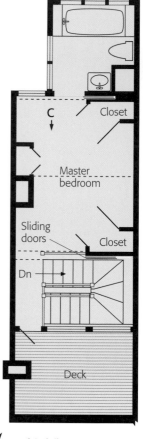

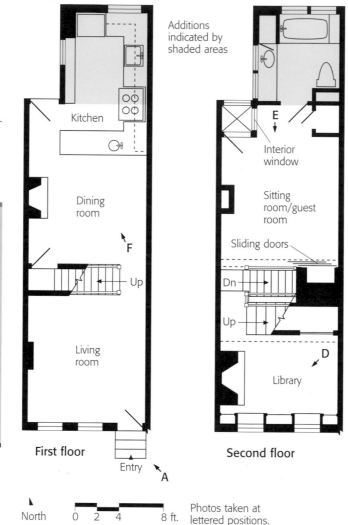

Additions indicated by shaded areas

Kitchen

Dining room

F

Up

Living room

First floor

Entry

A

E

Interior window

Sitting room/guest room

Sliding doors

Dn

Up

D

Library

Second floor

B

C

Closet

Master bedroom

Sliding doors

Closet

Dn

Deck

Third floor

North   0   2   4   8 ft.

Photos taken at lettered positions.

As we developed the building plan, we attempted to suggest rather than to define the living spaces.

durable, yet handsome enough to complement the formal spaces. The cabinetry is natural cherry; the countertops are Tennessee pink marble.

On the second and third levels, we enclosed each bathroom with walls topped with operable transom windows. These oversize windows allow exterior light and air into the house while permitting privacy in the bathroom.

The existing stair from the first floor to the second floor had been built against a partition wall and created a visual barrier between the front and rear of the house. We removed the wall and rebuilt the stair in the same location, extending it up to the new third floor. By leaving the stair open on all levels, we created easy circulation flow vertically and horizontally, and invited views from one level to another. On each floor, the stair provides a visual delineation between spaces without actually separating them from one another.

Sliding shojilike doors of Douglas fir with sandblasted glass can separate the library from the sitting room and can give privacy to the second floor (see photo p. 17). On the third floor, sliding wood screens with operable shutters work similarly to close the bedroom suite from the stair. Both sets of sliding doors used tandem roller hardware by Häfele (see Sources).

Rumford fireplace adds warmth to a cozy library. The architects chose to use a Rumford-inspired fireplace in the library because the fireplace's shallow profile takes up little space and throws lots of heat. Photo taken at D on floor plan.

Rumford Fireplace Is a Good Choice for a Small Library

Creative use of existing space included space-saving elements such as bookcases and storage cabinets built into the stair, two-sided corner closets in the bedroom and the incorporation of a Rumford fireplace in the library (see photo above). Rumford fireplaces, based on a 200-year old design originated in England by Count Rumford, are characterized by a relatively tall, shallow, straight-backed firebox with a small, rounded chimney breast. Rumfords are energy-efficient space savers and burn cleaner than traditional fireplaces, making them an ideal design for urban residential situations. After investigating Rumford-design techniques, we came up with our own version, incorporating liner components manufactured by Superior Clay Products (see Sources). The fireplace, detailed with painted-wood mantel and handmade tile, lends character and warmth to the room without sacrificing valuable space.

Tony Atkin, FAIA, is a partner in the firm Atkin, Olshin, Lawson-Bell Architects in Philadelphia, Pennsylvania.

SOURCES

Marvin Windows and Doors
Warroad, MN 56763
(800) 346-5128
www.marvin.com

Kolbe & Kolbe
www.kolbe-kolbe.com

Häfele
3901 Cheyenne Drive
Archdale, NC 27263
(800) 423-3531
www.hafele.com

Superior Clay Products
P. O. Box 352
Uhrichsville, OH 44683
(614) 922-4122

A Top Floor
with a Low Profile

ON A SUNNY DAY IN THE MISSION HILLS NEIGHBORHOOD OF San Diego, I stood in the middle of the street with my clients. Trying to describe their vision for a second-story addition to their small home (see photo bottom facing page) they pointed to a neighbor's second-story addition and said, "Anything but that."

Their Spanish-style house is nearly 80 years old, which for Southern California borders on ancient. To accommodate their growing family, they wanted to add a second story with three bedrooms and two baths (see floor plans p. 25). But to be in keeping with the established neighborhood, they also wanted to maintain the scale of the existing house.

The existing house suffered from an uninteresting layout, but it had several charming elements that were worth retaining: a trio of arch-top windows in one of the front bedrooms, an entry tower with small stained-glass windows, and an overall scale that fit well in the neighborhood.

A modest Spanish-style stucco home. The original single-story house fit the neighborhood but not the owners' growing family. Photos taken at A on floor plan.

RIGHT Smaller rooms go in front to control scale. This small bedroom presents a narrow profile to the street, with the rest of the second story recessed behind the flat roof over the living room to make the house look smaller. Photo taken at B on floor plan.

BELOW Keeping a low profile. Sloping the ceilings toward 7-ft. 6-in. top plates on exterior walls of the master bedroom allowed the tile roof above to lie low. Photo taken at D on floor plan.

Maintaining the Scale Makes All the Difference

I can always tell a modern Spanish-style house by the height of the exterior walls; new houses always seem to grow a bit too tall as we satisfy the desire for high ceilings. In this project, we maintained the traditional scale by carefully arranging both the exterior elevations and the location of new interior spaces. Most notably, we lowered the second-floor top plate to just 7 ft. 6 in. rather than the standard 8 ft.

Backyard becomes a courtyard. Wooden railings on the master bedroom's balcony are less formal than the wrought-iron railings that are in public spaces. Photo taken at H on floor plan.

To prevent the new rooms from feeling too squat inside, especially for our 6-ft. 4-in. tall homeowner, we created tray ceilings in the upstairs bedrooms. The ceilings slope up from the lowered top plate, then level out either at 8 ft. 6 in. at 8 ft. 9 in. (see photo bottom p. 22). This strategy yielded more interior height without adding exterior mass.

Careful arrangement of the floor plan can minimize exterior scale. For instance, modern rooms tend to be wider as well as taller, which translates to a bigger exterior mass. To keep the new front elevation from becoming too bulky, we positioned a small (children's) bedroom at the front (see photo top p. 22). The big master bedroom is behind the children's rooms, not only to orient it toward the private space of the backyard (see photo p. 23), but also to reduce the scale of the front elevation.

Structural Problems Artfully Solved

We faced several structural challenges in this remodel, and contractor Jim Broutzos of Wildwood Development in La Jolla handled them beautifully. The first requirement was to create internal shear walls to comply with California's strict seismic code.

To our dismay, we found the interior walls were made of 2x3 studs. These smaller studs, common in the older homes built in the area, are inadequate either for taking shear loads or for supporting a second story. We widened the existing sole plate and sistered new 2x6 studs to the original studs.

The new wall framing also had to be taller to support the second-story joists. But adding a kneewall would have created a hinge joint between floors, which is unacceptable in seismic regions. So we balloon-framed the new studs, notching the existing top plate to let the studs rise to a new top plate under the joists. Finally, these interior walls were sheathed with ½-in. plywood to create the shear panel.

Retrofitting the shear walls and pouring deeper footings below them required major demolition. But it had to be done without removing some first-floor wall finishes we wanted to save, including ceramic tile in a bathroom and coved plaster ceilings in the living room and dining room. To accomplish this, Jim's crew tore out only one side of each wall that was to be strengthened, working carefully from the side without tile or coving.

Another challenge faced us with the chimney. When you're adding a second story, any chimney within 10 ft. of new construction needs to rise at least 24 in. above the height of the addition. With the seismic requirements in Southern California, our local building department frowns on changes to existing fireplaces. Instead, they prefer that old, unreinforced masonry fireplaces be taken down entirely, no matter how valuable or coveted the finish on the fireplace might be.

Because the existing fireplace had original Batchelder tiles (found occasionally in Craftsman-style homes) decorating the fireplace surround, we tore down the chimney only as far as the smoke chamber and rebuilt it with a lightweight metal flue. We then built a wood-frame stucco wall around the metal flue. The new chimney is less top heavy and more stable in the event of an earthquake, and it's also high enough to allow adequate draw. The original tiles were left in place.

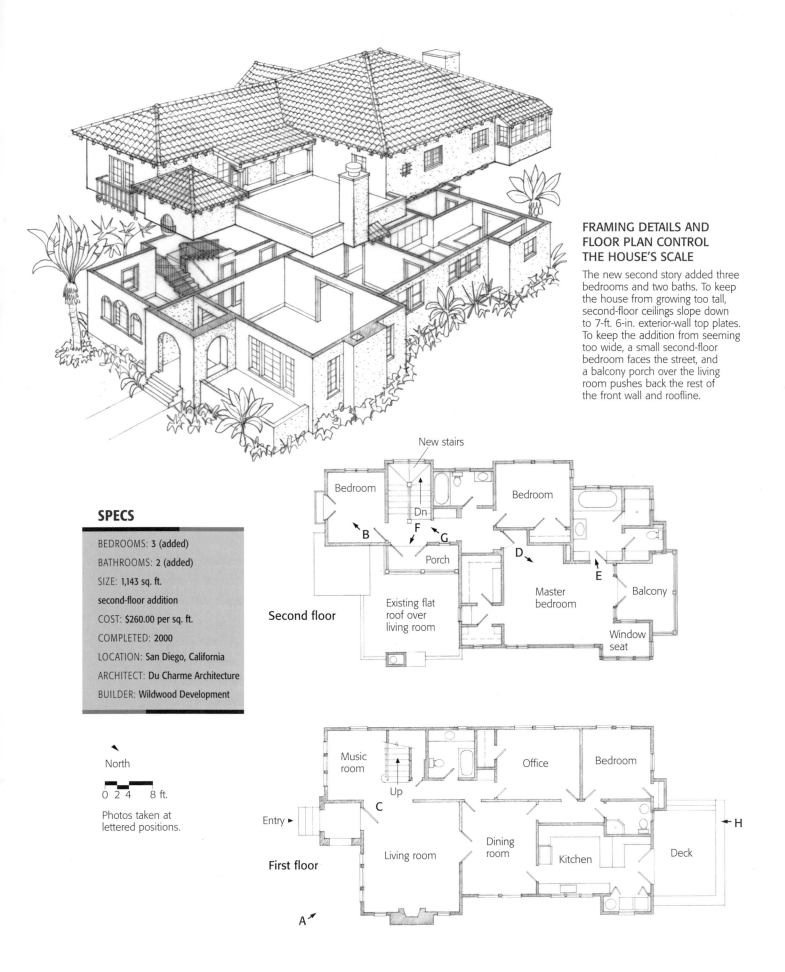

FRAMING DETAILS AND FLOOR PLAN CONTROL THE HOUSE'S SCALE

The new second story added three bedrooms and two baths. To keep the house from growing too tall, second-floor ceilings slope down to 7-ft. 6-in. exterior-wall top plates. To keep the addition from seeming too wide, a small second-floor bedroom faces the street, and a balcony porch over the living room pushes back the rest of the front wall and roofline.

New stairs

Bedroom

Dn

B

F

G

Porch

D

E

Bedroom

Balcony

Second floor

Existing flat roof over living room

Master bedroom

Window seat

SPECS

BEDROOMS: 3 (added)

BATHROOMS: 2 (added)

SIZE: 1,143 sq. ft.
second-floor addition

COST: $260.00 per sq. ft.

COMPLETED: 2000

LOCATION: San Diego, California

ARCHITECT: Du Charme Architecture

BUILDER: Wildwood Development

North

0 2 4   8 ft.

Photos taken at
lettered positions.

Music room

Up

C

Office

Bedroom

Entry ▶

H

First floor

Living room

Dining room

Kitchen

Deck

A

ABOVE **Windows light a dark wall.** The landing window helps to light the deep blue walls of the stairwell. The wrought-iron handrail matches the railings outside. Photo taken at G on floor plan.

LEFT **Adding stairs changes the entry hall.** A bedroom became a music room, and the arched windows became part of the public space downstairs. Photo taken at C on floor plan.

The existing house suffered from an uninteresting layout, but it had several charming elements that were worth retaining.

Interior Details Need to Match the Style

Except for strengthening the shear walls, the only major work on the first floor was building in a stairway where a front bedroom had been, with the leftover space becoming a music room (see photo top left). We removed a wall that separated the living room from the bedroom to give us enough space for the stairs and music room. With this wall removed, the main living spaces now enjoy views through the arched period windows that were previously hid-

den in the front bedroom. The staircase's mission-style wrought-iron handrail (see photo top right) matches the railings on the front and side balconies. The stairway is lighted by a window at the middle landing and, on the second level, by French doors that open onto a small porch (see photo top facing page).

The clients selected the deep, rich colors for most of the spaces with only a little guidance from my office. Although the color samples seemed a bit daring at first glance, the colors were perfect once the

LEFT A private porch. French doors light the stairwell and hallway and lead to a porch over the flat-roofed living room. Photo taken at F on floor plan.

BELOW The master bathroom, painted in traditional white, was placed in a back corner of the house, which allowed for higher exterior walls. Photo taken at E on floor plan.

walls were painted. Well, there was that "Pepto-Bismol pink" in a daughter's bedroom that had to be toned down. But the client is a gastroenterologist, so maybe the pink was appropriate. And when the study walls were painted a deep blood red, the client asked for a more conservative hunter green. The bathrooms were finished with a traditional white (see photo above).

Laura Du Charme Conboy, Principal of Du Charme Architecture in La Jolla, California, has been practicing residential architecture for fifteen years.

Elevating the Shingle Style

S OMETIMES, IT'S THE INVISIBLE PARTS OF A HOUSE THAT EXERT
the strongest influences on the way a house looks. For example,
the site Lola and Chuck had chosen for their house was on a hill-
side that sloped toward views of Rhode Island's Narragansett Bay. The bay
could be seen only from the highest point on the lot. But unfortunately,
the upper part of the slope was also the only place where the septic system
could be installed.

Grand gables in front, lesser gables in back. A sheltered entry and a pair of large gables play down the height of the house from the front (photo taken at A on floor plan).

THE RIGHT ROOMS GET THE RIGHT VIEWS

The round window wall on the first floor provides panoramas from the living room and family room, while the open kitchen shares the view. The dining room, used mostly in the evening, was placed in the rear of the house. The master bedroom gets its views from a recessed deck and balcony.

SPECS

BEDROOMS: **4**

BATHROOMS: **4½**

SIZE: **4,400 sq. ft.**

COST: **n/a**

COMPLETED: **1995**

LOCATION: **Southern Rhode Island**

ARCHITECT: **William Burgin Architects**

BUILDER: **Robert Morin**

North

0 2 4   8 ft.

Photos taken at lettered positions.

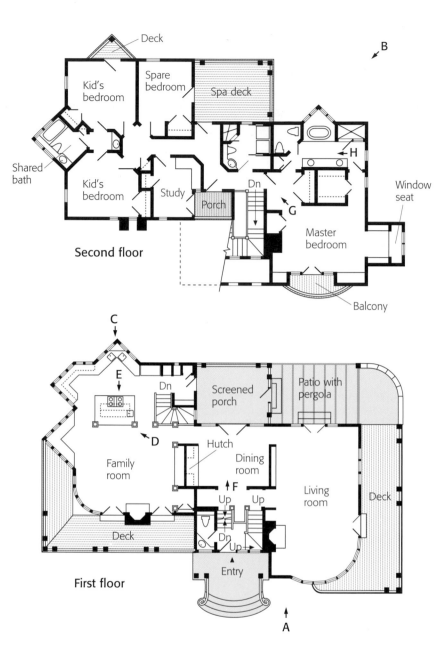

Second floor

First floor

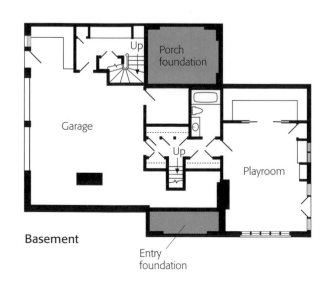

Basement

The next best place for the house was midway downslope. So to raise the house high enough to catch the views, I needed to put the first floor of the house one story above grade. This placement gained the height necessary for the views but simultaneously posed another problem: how to keep the house from looking too tall. In this case, the solution turned out to be an arched portal entry that engages the house midway between the basement level and the first floor (see photo pp. 28–29). This elemental shape works as a visual anchor, tying the house to its site. On the north side, hidden from the street, the garage conveniently tucks into the partially buried basement level.

Symmetrical Asymmetry

The extra height of the house is handled visually by two massive gables, one stepped back from the other. Each roof cantilevers over an open porch with a curved wall of glazing, giving the house a loose sense of symmetry. But that symmetry is altered subliminally by the other shapes and by the window arrangements.

The south gable is dominated by curved shapes starting with a bow roof above the second-floor balcony and ending with a curved window wall on the first floor. The curves get progressively wider and move farther off center as they cascade down.

The north gable is punctuated by a pair of chimneys that rise from a singular mass veneered in stone. The windows on this gable reflect the asymmetry with more glazing toward the center of the house. Two windows are also sandwiched between the chimneys. Inside, the lower window ends up in an unlikely spot directly above the fireplace.

A Different House from the Back

With the rising grade, the uphill side of the house couldn't support the same massive shapes visually, so I scaled back the volumes (see top photo below). From the back, the roof over the south porch is revealed to be mainly support for an ample dormer in the master bedroom, giving that part of the house the shape of a two-story Cape.

This elemental shape works as a visual anchor, tying the house to its site.

ABOVE **The rooms at the back of the house** don't have the spectacular bay view as the rooms in front, but they offer the homeowners privacy and a sense of outdoor living inside with the large screened porch, second-floor balcony, and terracing into the landscape.

LEFT **Bathing in the bay.** An angled bay on the back of the house is the perfect niche for the tub in the master bath. Photo taken at H on floor plan.

Kitchen island surrounded by bays. Triangular bump-outs create dramatically angled interior spaces. Inset photo taken at C on floor plan. Photo above taken at D on floor plan.

I tucked a screened porch into the two-story corner between the two house sections, turning the north part of the house into a saltbox. The line of the porch roof extends into a pergola that covers a stone

patio and helps to tie the two house sections together.

Two angled bays accent the back of the house. Above the pergola, one bay forms a tub alcove in the master bath (see photo bottom p. 31). A larger bay on the first floor creates a light-filled niche with views of the garden from the kitchen sink (see photos inset facing page and above).

Shared views. The kitchen opens into the family room with views of the water beyond. Photo taken at E on floor plan.

Cabinetry above the kitchen island was kept high enough for a person cooking at the stove to be inspired by the views.

Distributing the Views

Lola and Chuck had strong priorities about which rooms needed the views. The family room, where they would be spending most of their time, occupies the north section of the house with the best view of the bay (see floor plans p. 30).

An open kitchen shares those same views. Cabinetry above the kitchen island was kept high enough for a person cooking at the stove to be inspired by the views as well. And an informal eating area built into the angled bay above the garage offers diners bay views while they sip their morning coffee.

The dining room, which is used primarily at night, didn't need to have those views. So I put it in the back of the house off the screened porch. Large French doors can be opened, expanding the dining room into a bigger space. The living room, which literally rounds out the first floor, reminds me of a promenade deck on a cruise ship with glimpses of the bay in the distance.

Spectacular Bedroom Lookout

Upstairs, the master bedroom and bath occupy the entire south section of the house. The bedroom peeks out around some tall trees to reveal a spectacular view of the bay from a recessed deck and the curved "Romeo-and-Juliet balcony," as Lola calls it (see photo facing page).

The north section of the house is split into four rooms: two children's bedrooms, a guest bedroom, and a small study. The children's bedrooms share a bathroom via a common hall that has an additional vanity and sink.

The study opens onto a small porch hidden in an alcove where the two sections of the house connect. This tiny porch is the perfect spot for contemplation or for a glass of wine while watching the sunset over the bay.

Bedroom with a balcony. The master bedroom shows off all the intersecting rooflines that make the exterior so interesting, but the focal point is the recessed deck and the curved balcony. Photo taken at G on floor plan.

Finish Details Are Simple but Elegant

The detailing on the interior was kept fairly understated. Simple painted moldings throughout the house take a subordinate role to the clients' collection of Craftsman-style furniture. And a built-in hutch made in this same style tucks into an alcove in the dining room (see photo above).

Outside, the finish details were also kept simple, with flat uncluttered moldings around all the doors and windows.

The deck railings are a bit more decorative. We made the posts more massive than

necessary, and then devised a simple railing pattern that turns up on the exterior decks and the interior stairway.

William L. Burgin, who lives and works around Narragansett Bay, has an architectural firm in Newport, Rhode Island.

Making Room

A S MY CLIENT AND I NEARED THE HOUSE HE AND HIS WIFE were about to pour their hard-earned money into, he pulled the car to the curb and shifted into park. Something worried him. "About this place," he said. "It really is a beautiful lot. Unfortunately, the house stinks."

He drove on and soon parked in front of a tall hill crowned by an extraordinarily ordinary little house (see photo below). His assessment of the house had been correct; his worry, however, was misplaced. The lot—in a pleasant, older Minneapolis neighborhood filled with mature trees—was on a hill overlooking a lovely stream. Unfortunately, bad remodeling had reduced the house to a suburban saltbox out of scale with surrounding houses. Also, the two-bedroom house was too small for my client's family.

Fortunately, there was plenty to work with. The back half of the house had not been ruined by remodeling. The kitchen, the dining room, and the family room downstairs along with a bedroom and a bathroom upstairs (see floor plans p. 40 left) were all in good shape. In addition, the foundation and general construction of the house were in great condition.

ABOVE From cracker box to family home. Plain in appearance, small in scale and inadequate in size, the existing house needed more room and more style.

FACING PAGE The remodeled house, which adopts the Prairie Style, has ample room, better light and a scale in proportion to its neighbors.

Second-floor plan before

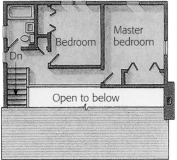

Bedroom

Master bedroom

Dn

Open to below

MORE AND BETTER UPSTAIRS SPACE

The original house had two bedrooms and one bath upstairs. By bumping out the front, 32 sq. ft. was added to the main floor and 232 sq. ft. to the second floor. Raising the roof height opened up the front rooms. At right is a front-to-back section through the front door.

Steel I-beam

Second-floor plan after

Bedroom

Dn

Bedroom

Master suite

A structural overhang and "energy trusses."

The 3-ft. eave overhang is an extension of the energy trusses, which have a raised heel that leaves room for 12½ in. of insulation above the exterior wall.

Air baffle

R-44 insulation

Raised heel

Continuous soffit vent

6-mil vapor barrier

Airflow

First-floor plan before

Family room

Kitchen

Dining room

Living room

Up

Trusses put most of the loads on the front and back walls.

A new wall along the front of the house and stub-wall extensions above the center and rear walls support the roof trusses. Other structural components include second-floor joists of laminated-veneer lumber and a new post and footing in the basement.

New truss roof

Diagonal brace

Stub walls

3-ft. overhang

New LVL floor joists

First-floor plan after

Section shown above right

Family room

Kitchen

Dining room

Music room/entry

Up

Living room

New porch

Steel I-beam

Post and footing

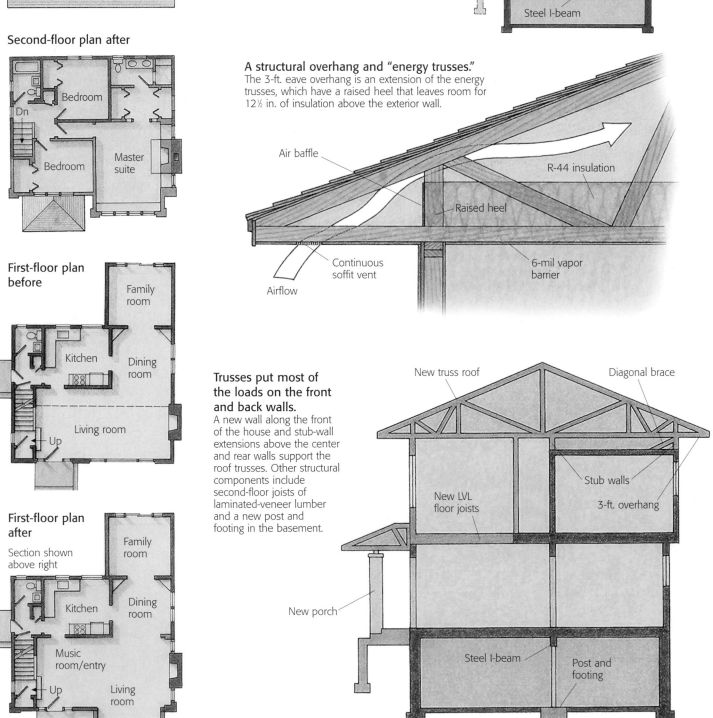

Tearing It Down Was Too Expensive

Because there was so much worth saving, it didn't make sense to demolish the house. So my client, Nick Smaby (who also happens to be my boss), and I hit the drawing board. It appeared that the only practical way to enlarge the house, given zoning restrictions, was to remove the existing roof and add a master bedroom and child's bedroom above in the front.

The main-floor living room also was a bit too small. Initially, we thought we would add up to 5 ft. across the front of the house; the addition could rest on the protruding tuck-under garage. The walls of the garage were large enough to act as foundation walls for any work above.

The idea of bumping out the front proved problematic with the local zoning department, however. At first, they refused to let us add anything to the front. After careful review, we realized that the site was on a curving street. The way local zoning ordinances read, the curve was sharp enough to consider this property as a corner lot. Because of this new status, the zoning department allowed us 3 ft. of addition.

Ultimately, 32 sq. ft. was added to the main floor; an additional 232 sq. ft. was added to the second floor by raising the roof and enlarging the house in front. Although adding conservatively to the square footage of the house, this approach transformed the dumpy one-story saltbox into an appropriately tall two-story building (see photo p. 39).

Structural Issues Dominated the design

When adding second-story space, the big question is always how to transfer new loads to the foundation. The easiest solution in this case was to have the new trusses span from exterior wall to exterior wall. Unfortunately, the old top plate on the second-floor walls was at 7 ft. 6 in. (see floor plans top right facing page). The new master-bedroom ceiling we wanted would be 9 ft. tall.

Our solution was to build a new wall to the correct height in front and a stub wall above the existing top plate in the center and rear of the house to make up the difference. To prevent the stub wall from rotating under the roof load, we had to brace it solidly back into the existing ceiling joists. The stub wall effectively became a vertical site-built truss (see floor plans bottom right facing page).

A thorough investigation of the foundation suggested that it was adequate for additional loads. Perhaps this was because the house originally had some second-floor space. However, a steel I-beam running side to side down the center of the basement was too small for additional loads (see floor plans bottom right facing page).

We kept most of the roof load off the center beam by clear-spanning from plate to plate. But we still had to match the depth of the existing second-floor joists. And because the new floor span from the center of the house to the front wall was greater than the span from the center of the house to the back wall (see floor plans bottom right facing page) the new joists would have to be longer and stronger.

To keep the joists the same depth, we used laminated-veneer lumber. This allowed us to match the existing joist depth, span a greater distance, and have the necessary strength. Then, instead of enlarging the center beam, we simply added a footing and post at midspan.

When adding second-story space, the big question is always how to transfer new loads to the foundation.

3-ft. Overhang Calls for Special Trusses

This neighborhood is full of two-story Prairie style homes, which were our design inspiration. Added height would help the home fit in better with its neighbors and take better advantage of the narrow, hilled site. If anything, we realized the house might get too tall for its width.

To offset this effect, we included a 3-ft. roof overhang all around. With transom windows near the soffit, the only way to get an overhang this large was to use energy trusses. In Minnesota, energy trusses are gaining in popularity.

An energy truss is created by increasing the height of the truss over the exterior walls (see drawing p. 40 middle right). Local codes dictate R-44 attic insulation, which suggests a 14-in. total heel height and allows 12½ in. of high-density batt insulation and a 1½-in. airspace. With the energy truss, the entire R-44 can be placed to the outside face of the stud wall, allowing for a much warmer and energy-efficient wall-to-ceiling detail. A sheet of cardboard baffle material (a Windwash air dam) was stapled to the outside face of the studs to prevent the air that vents through the soffit from blowing the insulation out of place. When insulation stays put, interior warmth doesn't escape into the attic. Eaves stay cold and well ventilated, which helps to prevent ice damming.

Energy trusses also provide a more structurally sound overhang, a serious consideration with the large 3-ft. soffits we designed for this house. Also, the energy-truss system allows for taller windows because the soffit is above the top plate. However, an energy truss makes a house appear taller, something I had to keep in mind when designing exterior proportions and details.

Inside and Outside, Prairie Style Details Unify the Design

The exterior design concentrated on ensuring proper proportions. Horizontal bands of trim and window banks in addition to a hip roof with large overhangs bring the house into comfortable balance (see photo p. 39). The window banks, with transoms, also take advantage of the glorious view.

We tried to stay true to the design concept in the interior trim work. In this area of the country, older houses, especially Prairie School and Arts and Crafts homes, used white-oak millwork. It is plentiful locally, and it is a durable, open-grained wood. The most stable, beautiful cut for white oak is quartersawn. We decided to use quartersawn, but cost nearly prevented its use. Harry Jensen from our shop suggested we resaw ¾ boards into ⁹⁄₁₆-in. trim. With a two-part casing (casing and backband), the proportions of resawn trim are pleasing (see photo facing page).

We saved the existing white-oak flooring in the living room, patching in at the new addition and resanding the floor. In the new master bedroom, we installed new white-oak flooring.

Todd Remington wrote this article during his tenure as an architect with the Choice Wood Company in Minneapolis, Minnesota.

Masterful touches for a master suite. From the bathroom through a short, arched hallway that contains his and hers closets, the new master bedroom is lit by a wall of tall windows with transoms and is trimmed out in quartersawn white oak to complement the Prairie style remodel.

Uncramping a Cottage

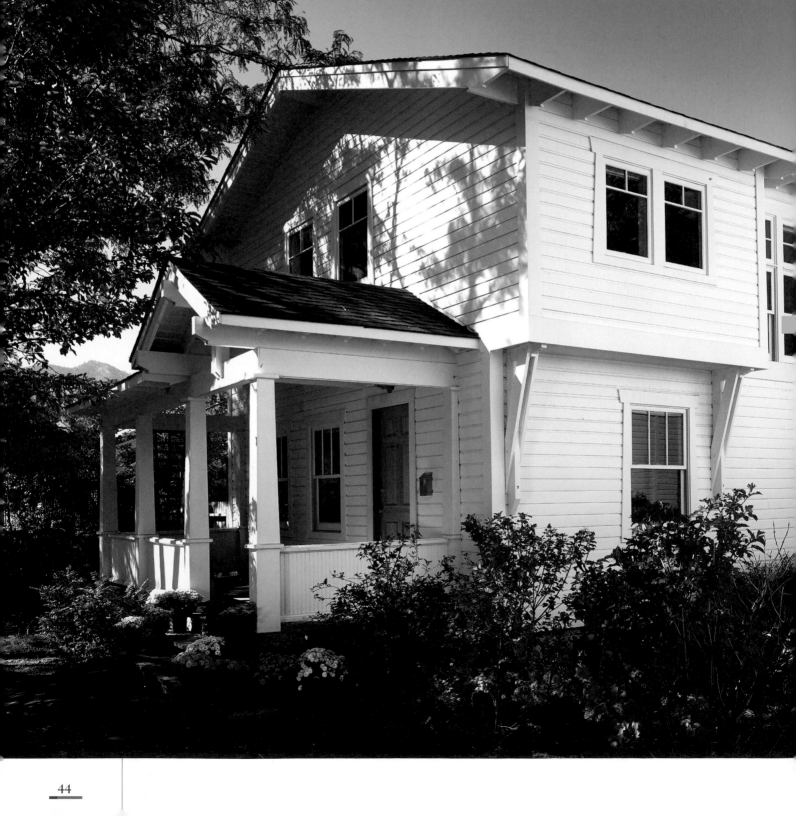

I T WOULD HAVE BEEN A STRETCH TO CALL THE LITTLE PLACE ON
Seventh Street an architectural gem. Undersized and unchanged since
its construction in the early 1920s, the single-story cottage was also
unsuited to contemporary living (see photo above). But to Sheila
McCullough and Peter Post, who had long been hunting for property in the
onetime mining boomtown of Boulder, Colorado, it was exactly right. The
800-sq. ft. cottage, still owned by the family that built it, was the only
unrenovated house on the street.

I've known Sheila since childhood, and I looked forward to working
with both her and Peter. Peter's background as a carpenter in New York City
and his desire to be the general contractor on his own house looked like big
advantages. Even so, I signed on as their architect knowing that challenges
lay ahead. The plan was to create a new house, nearly three times the size
of the original, on a site hemmed in by its neighbors. We could build in
only two directions: up and back. Zoning rules that preserved the cozy and
historic appeal of the neighborhood made the job even tougher by limiting
building height.

A new second floor and backyard addition
transform a tiny, outdated cottage (above) into a
mostly new home consistent with its neighbors
(facing page). Photo facing page taken from A on
floor plan.

Working under strict local
codes that limited building
height, the author found a way
to coax a spacious, three-bedroom
house from an 800-sq. ft. cottage.
A second story plus a 350-sq. ft.
addition into the backyard nearly
tripled the size of the original
structure. A hidden steel I-beam
that carries much of the
second-floor load also keeps
the living room free of partitions
and support columns. Photo
taken at A on floor plan.

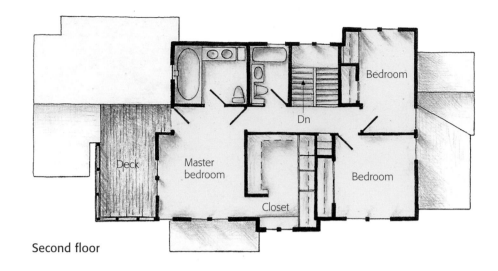

Second floor

*Another
advantage of the
floor plan is
that the natural
circulation flow is
along the east/west
wall, so no one
has to cut through
usable space on
the way to another
part of the house.*

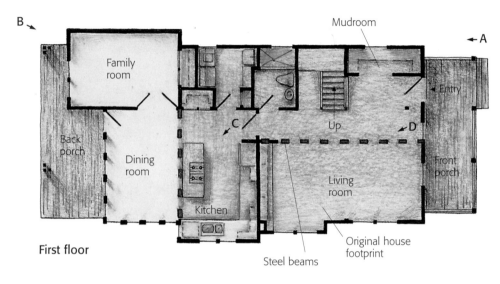

First floor

It didn't take long to realize that most of
the old cottage would have to go. I preserved outside walls on the front and sides
of the house, as well as the original first-
floor deck. Everything else was razed. New
living space came by way of a new second
floor and a 350-sq. ft. expansion into the
backyard, the one area where we were
unconstrained by setback requirements. The
backyard was private, with room for a new
garden. The plan looked like it would
work—providing we used interior spaces
efficiently and solved some tricky framing
and construction problems brought on by
the height restrictions (see sidebar p. 49).

Devising a Floor Plan That Opened Up Cramped Spaces

With a new house width of 26 ft. 6 in., the
best approach seemed to be creating as few
rooms as possible on the first floor so that
the space would seem larger and more
open. I used two dividing walls to organize
the space. The first crossed the width of the
house at about its midpoint (see floor plans
above), dividing the living room from the
kitchen/dining area. The second wall ran
the length of the house, sectioning off the
northern one-third of the first floor for sec-
ondary uses.

Kitchen and dining areas are separated by a large work island, which provides some division between the two spaces but also allows an overlap of uses (see photo right). Both areas share the generous number of windows and views of the backyard and garden to the west. Single-pane French doors are used to separate the family room from the kitchen; they give some privacy when needed without cutting off light or detracting from the area's open feeling.

Another advantage of the floor plan is that the natural circulation flow is along the east/west wall, so no one has to cut through usable space on the way to another part of the house. This basic approach changes the whole feel of the house. With its six equally sized rooms, the original cottage felt like so many little compartments. The expanded house has an airy, unconstrained feel to it.

It's easy to check on the garden. Big casement windows along the west and south walls of the dining area make the garden easily visible. Photo taken at C on floor plan.

Porches and Windows Are a Transition from Inside to Out

With interior floor space at a premium, both windows and porches were a good way to connect interior spaces with the outside, making the building seem larger.

The front porch was essentially rebuilt in its original location. Visually, the front porch is a buffer between the house and the street, which is only 18 ft. away. The porch and the traditional distribution of simple, double-hung windows give the front of the house a solid, traditional look that is in keeping with the neighborhood.

The backyard gave us more possibilities. Columns, an open gable roof over the back porch, and a lack of railings contribute to a sense of greater freedom and informality. Outside dining is possible in good weather, extending the usable area of the kitchen/dining area where the family spends most of its time. Shade in summer makes the

back porch an inviting place to relax (see photo top p. 48). We wanted the new garden to feel like an exterior room. So the easily accessible porch with glassed doors and a generous number of windows are especially important here.

Because of blustery mountain winds, local codes limit casement windows to 2 ft. 4 in. in width. This fact complicated window layout at the rear of the house. To avoid having different sash sizes and to keep an even window spacing across the kitchen wall, we built up the framing between windows. Three studs instead of the structurally required two at each mullion made up the difference. Concerned that a wide trim board might warp, we used two smaller boards with clapboards between them.

Kitchen and dining areas are separated by a large work island, which provides some division between the two spaces but also allows an overlap of uses.

Getting Eye-Catching Details from Plain Dimensional Lumber

With the high cost of construction, it seems as though the aesthetic details that give a house its character tend to be the first things that are compromised. We were able to achieve quality details with some additional materials, but mostly from intensive, skilled labor provided by Peter. Still, budget was an issue, so we looked for ways to get the most out of the materials we had.

Virtually all the wood we used was stock, but certain details kept it from looking that way. Built-up porch columns got extra posts where only one was needed structurally. They are more interesting visually and no more expensive than trimming out a single post conventionally. Brackets under the projecting second-floor bump-out were not required structurally, but they do lend the house some nice detailing.

Matthew A. Longo is an architect in Cambridge, Massachusetts.

ABOVE A backyard made for expansion. The front and two sides of the existing house offered no opportunities for expansion, but a generous backyard easily accommodated a 350-sq.-ft. addition with a decidedly relaxed feel. Photo taken at B on floor plan.

ABOVE A hidden steel beam opens the room. Above the living room is a 22-ft.-long steel I-beam that carries second-floor loads. Photo taken at D on floor plan.

Inventive Framing Lowers Roof Height

PROTECTING THE SUNSHINE

To guarantee that new houses do not block sunlight for neighboring lots, Boulder building codes limit how far shadows may extend past property lines.

In Boulder, Colorado, robbing winter sunshine from a neighbor is about as popular as stealing his horse would have been 100 years ago. You won't, in fact, get a building permit if the winter shadow cast by a 30° angle from the eve of the house creeps more than 6 ft. past the property line. The city calls this regulation a "solar fence," and it meant the roof on this house could be no higher than 26 ft.

A structural ridge raises second-floor ceilings. In figuring out the second-floor and roof framing, there was one more consideration: Colorado's snow load. To get the roof pitch we needed (no less than 5½-in-12) and keep the ridge below the magic 26-ft. cutoff, the top of the second-floor plate was set at 6½ ft. Roof trusses or conventional framing would have taken away too much headroom. Instead, I opted for a structural ridge that didn't need collar ties to prevent lateral spread. The ridge is a 3-in. by 9-in. Microllam supported at each end by columns that transfer loads through the first-floor framing to the ground.

Steel I-beams carry maximum loads in minimum space. With all the internal load-bearing walls on the first-floor deck demolished, we needed to find a way of supporting the new second floor without adding a lot of new partitions or lowering ceiling heights. The answer was two 10-in. steel I-beams. The longest is 22 ft. and runs along the east/west centerline of the house. One end rests on a 9-in.-deep Microllam window header that transfers its load through wood columns to two concrete piers poured at the existing foundation wall. The other end is supported by a column that rests on a 3-ft. by 3-ft. concrete pier poured in the crawlspace. Second-floor joists are 11 ⅞-in.-deep wood I-joists attached to the I-beam with Simpson joist hangers (see drawing below). The arrangement gives plenty of strength with a minimum of height.—**M. A. L.**

¾-in. plywood subfloor

1½-in. by 8-in. wood plate

GAINING HEADROOM WITH STEEL

Steel I-beams carry second-floor framing with less depth than wood beams would have required. Wood I-joists are attached to the steel with Simpson joist hangers whose top flanges are nailed to a wood plate.

Simpson top-flange joist hanger

10-in. steel I-beam

Wood I-joist, 11⅞-in. deep

Adding on but Staying Small

THE WORD COTTAGE USED TO SUGGEST SMALLNESS. No longer is this true. Today, a country cottage can be a 4,000-sq.-ft. second home with a three-car garage and a lap pool. This cottage, however, lives up to the humble origins of its name. It is a small country dwelling north of New York City, built midcentury as a summer cabin and, alas, not built very well.

To gain space, the first floor was enlarged, and a low-profile second story was added. Photos taken at A on floor plan.

The house started life as a one-story cabin with a gable roof and three small rooms with no hallways between them—you had to pass through one room to get to another. At some point, a bedroom was added to the back of the house under a shed roof, making the layout that much more clumsy.

The current owner, my client, bought it as a summer home, too. But after years of forgivable neglect, the little cottage could not even live up to this modest task. It was truly tiny—even discounting for today's gargantuan

A cottage grows in size but not in scale. Built in the 1940s, this three-room house had quirky charm.

Was there a way to carve out more space without upsetting the balance between house and garden?

standards—and flimsy. On my first visit, the floor joists deflected under my weight.

Despite its shortcomings, the cottage was worth saving, and saving in more than spirit alone. The humble structure sat near the top of a slope accessible only by a footpath. It overlooked a clearing that my client had, over the years, transformed into a magnificent English-style garden.

My client, Berenice, is a gardener and a bicyclist (she does not own a car), and she did not want to destroy this intimacy. There could be no driveway, no massive deck, no towering two-story addition. Was there a

way, she asked, to carve out more space without upsetting the balance between house and garden? Was there a way to grow while keeping the cottage a cottage?

A Summer Cabin Short on Amenities

Big or small, it seemed clear that we were looking at a major renovation and an addition. Berenice wanted a large, open living room with a fireplace and views of the garden. She also wanted a master bedroom and guest bedroom, along with a conveniently placed second bathroom (the original bathroom remained off the kitchen). In short, she wanted space.

Along with adding more space, we would have to upgrade the existing structure and systems. A summer cabin can get by with things a year-round house cannot, and this house had barely gotten by. It was built without heat and remained that way until someone installed electric baseboards. Most of the house sat on concrete piers over a 30-in. crawlspace, except for an 8-ft. by 10-ft. basement underneath the kitchen and bathroom. This cinder-block basement was prone to flooding, and as a result, the water heater was rusted beyond repair. The plumbing and electrical systems were poor, the insulation was less than adequate, and the framing was lightweight. The 2x6 floor joists were on 20-in. and 24-in. centers, accounting for the bounciness of the floor.

So how do you gain space without destroying what made the property special in the first place? The obvious answer was to expand the first floor and to add a second story for the two bedrooms and bathroom. The enlarged, open living room would include a dining area and expand the garden view. It seemed simple.

Full story would be bigger but not better.
This sketch shows how a full second story would have added floor space, but the house would have towered over its surroundings.

The client wanted a second floor added onto this cottage, but she did not want the house to tower over the site. The author proposed two options: a full second story (see inset) or a half-story with a gable dormer in front and a shed dormer across the back. The latter option (shown in the larger sketch above, from A on the floor plan) was chosen because it created space but kept the roofline low and unobtrusive.

A Half-Story Keeps a Low Profile

Things that seem simple don't always turn out that way, though. The house was close to the setbacks on two sides of the property, and the other two sides were given over to gardens. This limited how much we could expand the first floor (see floor plans p. 59). Another problem was the second story.

Because the house is on a hill, a full-height second floor would tower over the property and overwhelm the garden (see sidebar above). This is precisely what we wanted to avoid: gaining space at the cost of erasing the cottage's unpretentious charm. It was simply too high a price.

The solution was a smaller second story. We raised the outside walls just 3½ ft. from their original height. As a result, the second

**A long dormer expands the
upstairs.** The low roof limited the
upstairs floor space, so a long shed
dormer was added along the back.
The overhanging rake shelters the
deck and makes the roofline more
interesting. Photo taken at B on
floor plan.

floor has a kneewall and sloped ceilings, but
a shed dormer along the rear of the house
helps to compensate for the lost space (see
photo above). The long dormer can't be
seen from the front; from the footpath that
leads to the front door, the roof profile
remains unobtrusively low.

Another part of the solution lay in the
large, hip-roofed dormer over the front
door. This big dormer creates an open
space, or clerestory, over the front door and
stairway (see photo facing page). At the top
of the stairway, an open hallway, or
"bridge," connects the two bedrooms (the
bathroom is in between). This big dormer
brought some much-needed light into the
staircase and to the front entryway, and it

opened the views of the surrounding garden. This summer house has no central air-
conditioning, so it was important to keep
the second floor as breezy as possible.

Below the first-floor addition, we built a
new concrete-block basement with plenty of
perimeter drainage. This would be the space
for a new gas-fired boiler, which would
replace the outdated and expensive electric
heat. It would also gain some room for storage. In the original section of the house, we
reinforced the first-floor joists and subfloor.
We shortened a particularly long beam span
with a couple of concrete piers, and we
reinforced the joists by sistering new 2x6s
onto the existing joists.

Natural light warms the front entry. The hip-roofed dormer over the front door illuminates the front entry and upstairs hallway. From the upstairs, the big window affords a sweeping view of the garden. Photo taken at D on floor plan.

There are so many windows and such dense foliage in the garden that in spring the living-room walls seem papered with leaves.

Using the Outdoors to Illuminate and Decorate

Most gardens are planted after a home is built. Because this project was an addition to an existing house, the gardens came first and determined much of the design. Lots of windows bring in light and provide views of the lush outdoor foliage.

With foliage like this, you don't need wallpaper. In the combined living and dining room, large windows and a bay-shaped bump-out provide sweeping views of the garden and porch. Photo taken at C on floor plan.

Garden Views Are Everywhere

Creating garden views was something we did everywhere we could. Windows surround the great room, which takes up most of the first floor (see photo above). There are so many windows and such dense foliage in the garden that in spring, the living-room walls seem papered with leaves. On the staircase, a circular window offers another

view. Upstairs, there is a small deck outside the master bedroom overlooking the garden, enough room for two chairs and a small table (see photo top facing page). When the weather is warm, the separation between indoors and out is hardly noticeable.

This connection extends to the front porch, which is a far more generous space than in the original cottage. Porches are

LEFT **The master bedroom gets a terrace.** A small deck over the dining area is just large enough for two chairs and a table. Photo taken at E on floor plan.

LEFT **A porch is just an outdoor room.** The deep, formal front porch takes full advantage of the spectacular garden view. The hip-roofed, templelike dining area has the proportion and scale of an interior room, without the windows. Photo taken at F on floor plan.

Hiding a Gutter behind Site-Built Molding

Instead of off-the-shelf gutters on the second-floor deck, the architect and builder made the gutter an integral part of the roof detailing. The roofing—a single-ply, torch-down waterproof membrane—wraps into a trough behind the molding. The decking and sleeper assembly, or duckboard, sits on top of the membrane and can be removed for cleaning leaves and other debris.

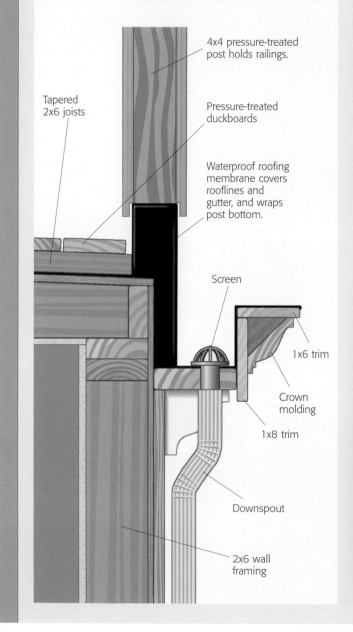

4x4 pressure-treated post holds railings.

Tapered 2x6 joists

Pressure-treated duckboards

Waterproof roofing membrane covers rooflines and gutter, and wraps post bottom.

Screen

1x6 trim

Crown molding

1x8 trim

Downspout

2x6 wall framing

A SUMMER CABIN GAINS PRECIOUS FLOOR SPACE

To give the original 500-sq.-ft. cottage (shaded area) more than twice as much space, the first floor was expanded and a second story was added. To gain space but maintain the low profile, a shed dormer was added along the back of the house.

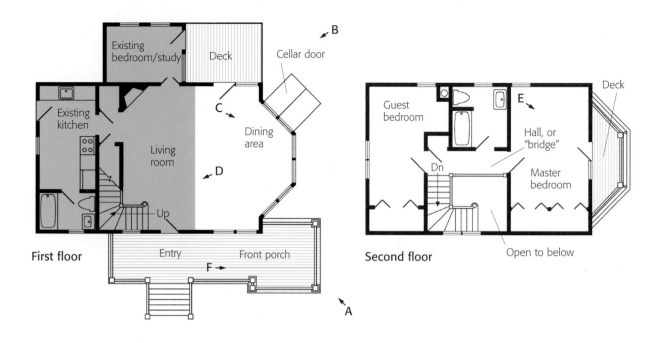

transitional spaces; they are part interior and part exterior. We tried to push this concept by adding the hip-roofed, templelike space for an outdoor dining table and chairs (see photo bottom p. 57). This area has the proportions and detailing of an interior space—it's almost like an extension of the living room, even when it's viewed from the inside. The new porch itself is deeper than the original, and the detailing is more formal, making it feel like an indoor space even though it is surrounded by plants and the open air.

Deep porches pose a problem—the deeper and more spacious the porch, the darker the adjacent interior space. This is where the clerestory came in handy once again. The amount of light that pours through the big dormer window counteracts the dark shadows cast by the porch.

Harry N. Pharr heads his own architectural firm in Warwick, New York. He currently serves as a member of the Village of Warwick Architectural Review board. The builder was Joe Nachtigal of Nightingale Construction in Warwick, New York.

The View Tower

I N BERNARD RUDOFSKY'S BOOK, *Architecture Without Architects*, there is a picture of a town where every house has a tower with a scoop to catch the breeze and bring it into the house below. The towers present a memorable skyline and signal a certain politeness and sense of sharing that goes lacking in many societies. Sylvia and John Quinn's house is our version of a similarly common architectural quest for a limited commodity—in this case, the view (see photo facing page).

The bones of the tower surround the breakfast table. A pair of beams and a solitary post engaged by the kitchen island emphasize the structure of the tower on the inside. Incidentally, the muntins of the big windows to the left of the table are reinforced with T-section steel bars to bolster them against the occasional hurricanes that pass over Rhode Island. Photo taken at D on floor plan.

When the flag flies, it means the owners are home. A three-story tower anchors the corner of a two-story addition to the rear of this simple New England cottage. The curvy railing on the top-floor balcony and the undulating fence were inspired by the waves on the nearby bay. Photo taken at A on floor plan.

The Quinns came to us with a turn-of-the-century house one block from the water in Jamestown, Rhode Island. It is a typical cottage, repeated with minor variations throughout the Northeast. The house faced the street to the north and blithely ignored the view of the water to the east. We meant to change that.

On the ground floor, our task was to renovate the kitchen in the rear wing of the house and to add a small dining space (see photo p. 60). On the second floor, we aimed to create a master-bedroom suite with a small sitting area. The top floor would be part outdoor balcony and part guest-room bunkhouse. On each level, taking advantage of the views was a top priority.

Building to the Zoning Limits

Zoning setbacks severely limited our options for lateral expansion. To the west, the existing wing was already on the setback line. To the south, we had 10 ft. to the setback and a couple of feet to spare on the east side. Vertically, however, we had another 20 ft. to go before we reached the 35-ft. maximum. Up became the direction for expansion.

At first we intended to keep the existing flat-roof rear wing as the ground-floor space and to add new rooms on top of it. But in the end, it proved unsalvageable because of its poor foundation. The mortar in the stone-rubble wall had turned to dust (a frequent problem on an island where beach sand used to be the standard material for mortar mixes). So the addition ended up as new construction, built in virtually the same footprint as the original kitchen. Only the corners of the tower (and the little deck) extend beyond the original footprint.

From the beginning, the 10-ft.-sq. tower symbolized the house's new direction. The form is skewed 10° (see floor plans p. 64) to catch better views and to distinguish it from the original house, both inside and out. The roof peak comes to within 2 in. of the 35-ft. height limit.

After some debate, we kept the tower corner posts on the first and second floors. They could have been eliminated with steel beams, but we thought it would be better to keep them and make the tower more apparent inside. It is further defined by beams in the ceiling and by a change in the direction of the flooring.

Simple Trim and Cabinets, with a Couple of Curves

The kitchen cabinets and other cabinetwork are deceptively simple. It's all birch-veneer plywood with flush overlay doors. A ¾-in. offset between the two rows of upper cabinets creates a shadowline at night when the overhead lights are on (see photo facing page).

We designed various cabinets to display pieces of glass and other objects Sylvia collects. The cabinets above the kitchen sink, for example, are separated by a tall window. This gap is spanned by a pair of glass shelves that serve as perches for artwork. During the day, light from the window illuminates the glass art, and at night, the pieces are lit by recessed low-voltage halogen lights trained on the glass shelves.

In stark contrast to the delicate glass shelves is the stainless-steel counter. It's a custom-made piece, 10 ft. long, that includes an integral sink and a cutout for the stainless-steel Thermador gas cooktop. At about $1,000, this counter is an affordable custom touch that really sets the tone in this kitchen. Simple stainless-steel door and drawer pulls from Haëffle (see Sources p. 64) carry the look of the counter to the crisp white cabinetry.

Plenty of seating near the kitchen. A round table occupies the base of the tower. In the background, operable windows sandwiched between the upper cabinets and the counter provide light and ventilation. Photo taken at B on floor plan.

SOURCES

Haëffle
(910) 889-2322

Section

A TOWER WITH A TWIST

Skewed 10° from the kitchen walls, the tower occupies the southeast corner of the addition. This twist sets it apart from the rest of the house and better orients the tower windows toward the view.

Bunk

Dn

Observation deck

Third floor

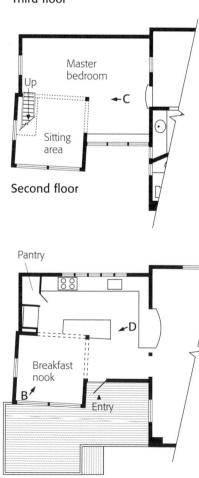

Master bedroom

Up

← C

Sitting area

Second floor

North

0 2 4   8 ft.

Photos taken at lettered positions.

Pantry

← D

Breakfast nook

B ↗

Entry

A ↗ **First floor**

A Ladder Leads to the Tower, While Waves Take Over the Fence

Tower access was a problem from the moment we conceived of the space. How do you fit a 3-ft.-wide stair with code-approved rise and run into a 10-ft.-sq. space? You don't. So we applied for and received a variance to allow a custom-built ship's ladder. The ladder works well (see photo top facing page) and takes up only a 2-ft.-wide slice of space.

John is a sailor, and the observation deck looks out to waters he has sailed since childhood. The hoisted flag means the owners are home. This sailing background and the proximity of the bay gave us the idea for the wavy railing on the observation deck (see photo p. 61). The idea spread to the kitchen backsplash (see photo bottom facing page). Then it got to the backyard fence.

The fence proved to be the most fun. The fence boards were installed at their full 5-ft. height. Then we met with a box of sidewalk chalk. After some lively, fairly irrational discussions about waves and many colored lines on the uncut fence, we decided on the final pattern, and our builders, Peter Medeirios and Steve Furtado, cut the fence boards and capped them with profiled trim. Later, the hedge clipper arrived and did his own version of the wave on the front hedge.

Maybe We Can Share the Views

In its final form, the Quinns' house remains a quiet cottage from the street, with no hint of the extensive renovations and the addition. From the east and from inside, however, the house is very different as it salutes the waterfront. The tower symbolizes this new direction as it pushes zoning to the limit with its modest 10-ft.-sq. footprint.

A ship's ladder leads to the observation deck. On the second floor, the tower is revealed by its structure and by the change in direction of the flooring. Here, the tower space is a sitting area adjacent to the master bedroom. Photo taken at C on floor plan. Photo by Charles Miller

A splash of cherry. A rich band of cherry wood wiggles across the wall, adding a sumptuous wedge of color to a refined wall of white cabinetry. Photo by Charles Miller

With minor or no zoning variations, waterfront towns could be dotted with narrow view towers—a modern version of Rudofsky's town. Each tower could take up a small slice of the skyline, offering enough view for everyone rather than the current situation where the person nearest the water builds to the maximum and consequently blocks everyone to the rear.

James Estes is a partner in Estes/Twombly Architects in Newport, Rhode Island. The firm specializes in residential work, both renovation and new construction.

It's just too small. Charming in its simplicity and located in a good part of town, this one-story house had been outgrown by its owners. Adding a second story solved the space problem, and using simple construction methods, including prefabricated trusses, kept the total cost to just over $100,000 (Canadian).

Adding a Second Story

A roof apron recalls the original proportions. Strong diagonal lines drawn by the 12-in-12 rake boards at the gable ends help to break up what would otherwise be a top-heavy facade. The lower roof continues across the front and back of the house, sheltering the windows and preserving the original roofline.

I IT WAS A WONDERFUL, PREMATURELY WARM DAY AT THE beginning of March 1994 when I first met Paul and Letizia Myers to discuss adding a second story to their house (see photo bottom facing page). Both of their children were in their teens, and the house was beyond feeling cramped. A second story would give Paul and Letizia a master suite, a room for each child and another bathroom (see photo top facing page).

That sunny March day had the kind of morning when tearing off the roof seems like the most natural and logical thing in the world. In fact, as Paul and I stood in the warm sun and looked at the roof he had repeatedly patched with elastomeric compounds, it seemed an unreasonable strain on anybody's patience to formulate a program, draw plans, and apply for permits.

In reality, the timing should have been perfect. The design could get done, and the plans drawn, in time to begin construction by late summer. August and September are the most reliably dry time of year in Vancouver.

But events foiled us. A strike at City Hall slowed the permitting process, and it was into November by the time we had approval to go ahead. Reluctantly, we shelved the project until spring. Then I met contractor Walter Ilg.

Walter makes a specialty of handling what he calls "the hard parts" of any renovation. I watched his crew remove and replace the foundation of my neighbor's house, and I was impressed with the expeditious way he handled the hard part of that one. So I showed him the plans for the Myerses' project. We agreed that the way to do it was to put up the new roof before taking down the old one. But we disagreed about timing. I had in mind the end of April. "Why wait?" he said. "It can rain anytime here."

It could do more than that, as were to find out. But on a warm Monday in March, almost exactly a year after my first visit with the Myerses, Walter and his crew started building scaffolding.

Prefab Trusses and Minimal Walls Help the New Roof Go Up Quickly

Walter's theory of framing is simple. You do the minimum necessary to get the roof on, throw a party, and then back-frame the rest. In this case the minimum was less than it might have been because the existing attic floor framing—2x8s on 16-in. centers—didn't have to be reinforced. Not that the job couldn't have been done the same way even if the existing joists had needed upgrading.

Preparing for the new roof. The crew begins construction of the new roof by excavating post holes in the old roof over the wall plate. On the left, a ramp for removing roof debris leads to a curbside Dumpster™.

Posts carry a wall beam. Well-braced with diagonal 2x4s, 4x4 posts rise from the holes in the roof to support a doubled 2x10 beam. Note the temporary flashings that are at the base of the posts. At the far end, the wall beam extends beyond the plane of the house to create a staging area for the roof trusses.

The new roof was also designed with minimums in mind: minimum cost and minimum delay. There would be no stick-framing; instead, factory-supplied trusses would carry the loads down the outside walls. Almost half of the trusses would be scissor trusses for the exposed wood ceiling over the stairs and in the master bedroom. The 12-in-12 pitch apron that forms the overhang at the gables and at the ground-floor eaves would be framed after the new roof was on and the exterior walls built.

To get the roof on, we needed just two bearing walls. But a continuous wall plate couldn't be installed without severing the old roof from its bearing. The solution was

to use posts and beams, and to frame in the walls afterward.

Based on the layout of the interior walls, Walter and I decided to use four 4x4 posts along each side of the house. The beams would be doubled 2x10s. In one place, one beam would have to span almost 16 ft., but any deflection could easily be taken out when the permanent wall was framed underneath it. As it turned out, there wasn't any.

So on Tuesday morning, with the scaffolding built, Walter's crew cut four pockets in the appropriate locations along each side of the roof (see photo top left). Then they secured the 4x4 posts to the existing floor framing and to the top plate of the wall below. They notched the end posts to fit into the corner made by the end joist and the rim joist. We were lucky with the intermediate ones; all of them could be fastened directly to a joist, notching the bottom of the 4x4 as required. None of the four intermediate locations was so critical, though, that the post couldn't have been moved a few inches in one direction or the other if necessary.

Walter used a builder's level to establish the height of the posts, and by Tuesday afternoon one of the beams was up and braced back to the existing roof (see photo bottom left), and the posts were in place for the other one.

At the same time, the rest of the crew was cutting away the ridge of the existing roof to allow the flat bottom chord of the common trusses to pass across (see photo top facing page). They were able to leave the old attic collar ties/ceiling joists in place, though, because the old ceiling had been only 7 ft. 6 in. I stopped by at the end of the day to inspect the temporary post flashings the crew had made with poly and duct tape. It had been another sunny day. By afternoon, however, thin clouds had moved in, and it was getting cold. The forecast was for snow.

The roof's ridge is cut away. With one beam in place, workers continue to chip away at the remaining ridge in order to make room for the trusses.

Snow and Rain Complicated the Job

The order for the trusses had been placed the previous week, with delivery scheduled for Thursday or Friday. But on Monday, while we were overseeing the lumber delivery, Walter let me know that he had called the truss company and promised them a case of beer if they delivered the trusses on Thursday and three cases if they got them here by Wednesday.

On Wednesday morning there was 8 in. of snow on the ground—and on the Myerses' roof. But the weather system had blown right through, and by 8 a.m. the snow was melting fast. Walter called to say he had sent two men to sweep the snow off the roof and that the trusses would be on site if the truck could make it out of the yard. At noon I arrived to see the last bundle of trusses being landed on temporary outrigger beams.

The rest of that day was spent finishing the beam and setting and bracing the trusses. Plywood laid across where the old ridge had been scalped made it easy for one man to walk down the roof supporting the center of the truss while two others walked it along the scaffolding.

The old roof comes down. With the new roof in place, the old one can come down. Next, the missing studs in the perimeter walls will be installed.

Even though it violated Walter's get-it-done roofing rule, I had the crew install the frieze blocking as the trusses were installed. By cutting the blocks with a chopsaw, you can ensure perfect spacing (even layout becomes unnecessary where framing proceeds on regular centers), and it's much easier to fasten the blocking this way than it is to go back and toenail it all afterward. Also, the soffit-venting detail I used with the exposed rafter tails required the screen to be sandwiched between two courses of soffit and stapled to the inside of the frieze block.

On Thursday another front brought wind and rain, which dispatched the last vestiges of snow but made a miserable day for the sheathing crew. Having to install the soffit, the screening, and the 2x4 purlins that tie all of the trusses together didn't speed things up. Nor did the four skylights. I didn't want the bargeboards done hurriedly, so to make things easier for the roofers we temporarily toenailed 2x4s on the flat to the trimmed ends of the rake soffits. That way the roofers could cut their shingles flush to the outside edge of the 2x4, and when the 2x4s were removed and replaced by the permanent 2x10 bargeboard and 1x3 crown, the shingles would overhang by a consistent 1¼-in. margin.

On Friday morning the roofers went to work on one side of the roof while the last nails were pounded into the sheathing on the other side. It didn't take long for them to lay the 12 squares we needed to make everything waterproof. Meanwhile, Walter and his crew were removing the old roof underneath (see photo bottom p. 69) and carrying it to the Dumpster in 4-ft. by 12-ft. chunks. I usually try to save the old rafters, but in this case I'm afraid I let the momentum of the job dictate the recycling policy.

By 1 p.m., true to his word and to long European tradition, Walter was tying an evergreen branch to the ridge and plates of cheese, bread, and sausage were being laid out on a sheet of plywood set up on sawhorses in the 26-ft. by 34-ft. pavilion that now occupied the top floor of the house. It might be a little breezy, as Paul said to me over a glass of wine, but at least it was dry.

A Roof Apron Prevents a Boxy Look

It took another three weeks to complete the framing and to do all of the picky work that's an inevitable part of tying everything together in a renovation. One detail, and an important element of the design, is the roof apron that encircles the house to break up the height of the building (see photo p. 66). The apron forms an eave along the front and back of the house. At the gable ends

A PREFABRICATED ROOF APRON

The horizontal roof apron that runs along the front and back of the house was assembled with 8-ft.-long, prepainted sections of rafters made up in the shop.

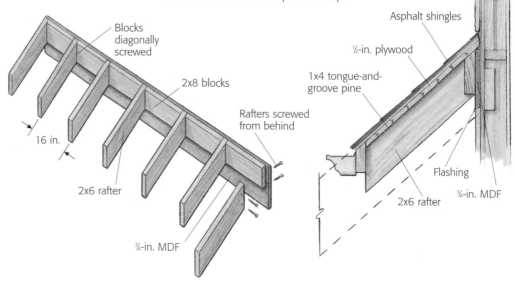

Blocks diagonally screwed

2x8 blocks

Rafters screwed from behind

16 in.

2x6 rafter

¾-in. MDF

Asphalt shingles

½-in. plywood

1x4 tongue-and-groove pine

Flashing

2x6 rafter

¾-in. MDF

the apron becomes a rake that rises to the peak of the roof, drawing long diagonal lines across what would otherwise be a tall, blank facade. The effect is of a 12-in-12 roof with 4½-in-12 shed dormers.

The apron has practical value, too, particularly at the eave, where it covers the top edge of the existing wall finish, providing an overhang to protect the ground-floor windows. If you're building outside the painting season, it's essential to get a coat of paint on everything before it's applied to the outside of the house, so we built as much as we could of this apron in 8-ft. sections in my shop (see drawing facing page). For example, the eaves consist of 2-ft.-long 2x6 lookout rafters screwed from the back to a 12-in.-wide strip of Medex (see Sources), an exterior-grade medium-density fiberboard that is gaining popularity for use as exterior trim here. Frieze blocks cut from 2x8s act as pressure blocks between the rafters. We prepainted these assemblies and the 1x4 tongue-and-groove pine that we nailed to their tops in our shop.

On site, the eave sections were installed and tied together with the prepainted 1x4s and 2x6 fascia. Then we snapped lines on the gable ends from the ridge to the eave lookouts to establish the line of the rake soffit (see photo above right). On this line, we toe-nailed a triangular bump-out, framed out of 2x10s, to the gable-wall framing. From the base of the bump-out we ran a 2x6 that acts as a rake trim board for most of its length and then becomes the last lookout rafter where it runs into the eave overhang.

We nailed preassembled and prepainted strips of soffit to the rake trim and to the gable bump-out. Made of tongue-and-groove 1x4s blind-nailed to 18-in.-wide strips of ½-in. plywood, the 8-ft.-long strips of rake soffit were pretty floppy until the 2x10 bargeboards went on.

Projecting the gable peaks out from the plane of the wall did more than provide solid support for the rake apron with its heavy bargeboard. It also created some visual interest and gave a little protection to the bedroom windows in the east wall. The peaks were finished with louvered vents and 1x4 bevel siding. These peaks make a nice big triangle of painted woodwork to balance the large areas of stucco.

We also ran a water table at the second-floor joist level. Besides its aesthetic contribution, this band covers the flashing protecting the top edge of the old stucco and makes a practical separation so that new stucco and old don't have to meet. Detailing woodwork so that stucco always has a place to stop and so that no one panel of it is too big makes the plasterer's job a whole lot easier.

Shaun Friedrich, who learned the stucco trade from his father and can tell without leaving his truck what a particular stucco is, when it was done, and quite often who did it, made a beautiful job of approximating the look of the original dry-dash finish. Dry dash is a labor-intensive stucco finish in which a layer of small, sharp stones is embedded in a layer of mortar. Shaun rendered a compatible finish for the upstairs walls by using a drywall-texturing gun to create the random, splattered look of dry dash. This substitution saved us $1,000.

Allocating the New Space

On the inside, Walter's crew was turning the 26-ft. by 34-ft. pavilion into a second floor with three bedrooms and two baths (see floor plan p. 72). The west end contains a master bedroom and bath. In the center of the house a hallway includes the existing stair, a bathroom at the north end (see photo top p. 73) and a balcony at the south end. Bedrooms at the east end complete the plan.

Bump-out and fascias support the rake soffit. At the gable end, a bump-out protects the upstairs windows and supports the tops of the 2x10 bargeboards. The 2x10s are borne by 2x6 fascias cantilevered past the roof-apron rafters. Note how a built-up water table makes a clean line between the old stucco and the new.

INJECTING VARIETY INTO A RECTANGULAR PLAN

Nooks, alcoves, skylights, and dropped ceilings all play their part in enlivening the plan.

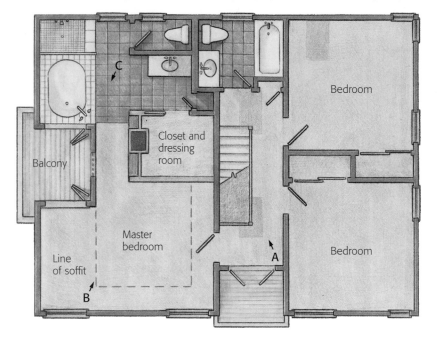

Balcony

Closet and dressing room

C

Bedroom

Master bedroom

Line of soffit

B

A

Bedroom

Daylight in the center of the house. A skylight over the centrally located hallway lights up the stairs, as well as the bathroom, by way of its generous transom. Photo taken at A on floor plan.

SOURCES

Medite Corp.
P. O. Box 4040
Medford, OR 97501
(541) 773-2522

The subdivision of the master-bedroom space to accommodate a walk-in closet and the bathroom was the most intriguing part of the design. I wanted the room to feel large and generously proportioned, but at the same time I wanted the different areas within it to be well defined.

The first division is between north and south. The bathroom, with its requirement for privacy, is on the north side; the bedroom is on the south. What separates them is not a wall but other subsidiary spaces: the walk-in closet and a small balcony (see photo top facing page).

Then there is the division between the main part of the room and the three 6-ft. deep alcoves along the west wall. Linked by their common ceiling height—7 ft.—the alcoves contain, from north to south, the shower/tub space; the balcony; and the bedroom-dresser area.

In addition to their common ceiling heights, the alcoves are further linked by large windows that open onto the balcony (see photo bottom facing page). These win-

dows can be folded back against the wall so that in nice weather the balcony is really a part of the bedroom.

The transparency of these linked alcoves to one another goes a step farther. On the bathroom side the shower is separated from the tub by a glass partition; on the bedroom side a window in the south wall lines up with the two windows to the balcony. Standing in the shower, you can look right through four transparent layers to the outside. In a small house, long views such as these foster a sense of spaciousness.

The ceiling in the master bedroom is an example of how to turn a technical problem to practical advantage. The decision to use trusses throughout for the sake of expeditiousness and economy meant that the ceiling could slope only at a pitch of 2-in-12 (the bottom chord of a 4½-in-12 scissor truss), and that skylight wells would necessarily be rather deep. Locating the skylight so that the slope of one side of the ceiling extends into the skylight well to meet the top of the skylight makes for a dramatic

LEFT **A balcony separates the bedroom and the bath.** Along the west wall, three alcoves with low ceilings have distinct functions. In the foreground, the shower and tub occupy the first alcove. In the middle, a small balcony overlooks the secluded backyard. In the distance, the third alcove provides space for the bedroom dresser. Photo taken at C on floor plan.

BELOW **The outdoors is nearby.** On the left, folding windows lead to a balcony off the master suite. On the right, a 7-ft. closet wall separates bedroom from lavatory. The sloping ceiling extends beyond the ridge to become part of the skylight well over the closet. Photo taken at B on floor plan.

light shaft that spills light all along the ceiling as well as down the wall (see photo right). It also didn't leave much room for error in the layout we had to do back on that raw day in March when Walter's crew members were swarming over the roof with snow in their hair and shinglers at their heels.

As for the low slope of the ceiling, we made it seem higher by holding the closet walls to a height of 7 ft. In the end the effect was everything we had hoped it would be. Letizia, who is Swiss and for whom I was trying to echo a wooden chalet ceiling, was not disappointed.

Tony Simmonds has been involved in building and woodworking for 25 years and operates DOMUS, a design/build firm in Vancouver, British Columbia, Canada.

Original half-Cape (photo right). A diminutive house built by the author's father in the 1960s was the starting point for an ambitious expansion project.
The original half-Cape remains the visual focal point. New living areas were added to the right and rear of the center structure. Photo taken at A on floor plan.

Expanding a Half-Cape into a Full-Blown House

J O AND BILL CRANE'S INVITATION TO REMAKE THEIR COTTAGE on Cape Cod was almost like getting an invitation to come home. I had known the Cranes, and the house, since I was a kid—in fact, my father had built the house for them back in the 1960s as a summer place. It was typical of a half-Cape, the simplest version of New England's oldest vernacular building type (see sidebar p. 79) And the Cranes' house reflected the usual strengths and weaknesses of this venerable style (see photo above).

My father, Paul, is a devoted student of the Cape. He had adapted some of the simplest traditional details to save both labor and materials. There was no exterior molding, only thin, square-section pine trim along the fascia and rake boards. A wide frieze board served as the head casing for the 12-over-8 double-hung windows. The rafter tails were square-cut to create the slight overhang typical of old Cape houses while eliminating the need for a soffit.

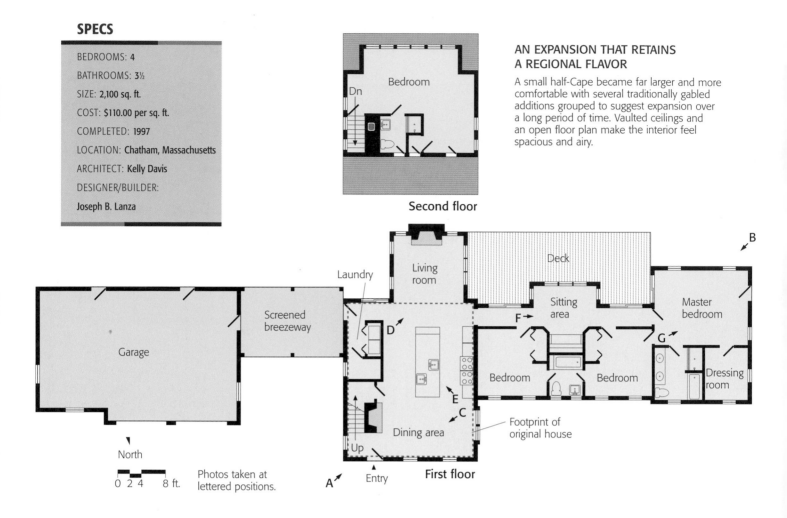

Bedroom

Dn

Second floor

AN EXPANSION THAT RETAINS A REGIONAL FLAVOR

A small half-Cape became far larger and more comfortable with several traditionally gabled additions grouped to suggest expansion over a long period of time. Vaulted ceilings and an open floor plan make the interior feel spacious and airy.

B

Deck

Living room

Laundry

Sitting area

Master bedroom

F

D

Bedroom

Bedroom

G

Dressing room

E

C

Footprint of original house

Dining area

Up

North

0 2 4 8 ft.

Photos taken at lettered positions.

A

Entry

First floor

Garage

Screened breezeway

The house needed a lot more light and roaming room to counteract long New England winters.

Over the years, the house had grown in the way that old Capes often did. In the early 1970s, a small saltbox garage went up, along with a 12-ft. by 18-ft. screened breezeway connecting the garage to the east side of the house. In the early '80s, a shed dormer was added across the back of the house to increase headroom upstairs. In all, the 1,000-sq.-ft. house had a pleasing simplicity and scale. But then there were the obvious negatives for a house that was about to graduate from summer place to year-round dwelling: a lack of natural light inside and cramped spaces, especially in the kitchen. The house needed a lot more light and roaming room to counteract long New England winters.

I started with a long list of what the Cranes wanted: four bedrooms, including two potential master suites; at least three bathrooms; separate living and dining areas; a two-car garage with additional boat storage; and an enormous kitchen that would be both a serious cook's dream and the true heart of the house. Soon afterward, reality caught up with us. At a total of 3,200 sq. ft., this first try was not so much an addition to the house as it was a complete engulfment. It was way too much house for the site.

Paring Down the Owners' Wish List Leads to a More-Open Floor Plan

We began distilling the wish list to its essentials. The big kitchen was sacrosanct. Bedrooms and baths were necessary to accommodate friends and family. But as to

LEFT A living room becomes a dining room. The largest room in the original half-Cape was a 21-ft. by 11-ft. living room that became this generous dining area. Photo taken at C on floor plan.

BELOW Massive island is the heart of the kitchen. A 48-sq. ft. island provides enough room for two sinks. The author poured the concrete counter in place. Photo taken at E on floor plan.

the rest, light, comfort, and a feeling of spaciousness were more important to the Cranes than square footage. So the idea of separate, formal rooms was thrown out. In its place, we substituted an informal arrangement of adjoining spaces (see floor plans facing page). The front door on the north side of the house opens into a corner of the new dining room—what had been the living room—next to the original brick fireplace (see photo above). This space adjoins the kitchen, where a 4-ft. by 12-ft. concrete-topped kitchen island (see photo right) stretches toward the living room and its own fireplace, 42 ft. from the front door. The whole seems larger than the sum of its parts because each area borrows space from both adjacent rooms and outdoor spaces.

ABOVE Capes have a charm all their own. Broad and low to the ground, the Cape Cod house remains a signature New England structure with a warmth matched by few other buildings.

ABOVE A midhall reading nook was created with a little space borrowed from a bathroom and a little taken from the deck. At the far end of the hall is the master bedroom. Photo taken at F on floor plan.

RIGHT Gabled additions suggest this cape grew over time.

The Cape Cod Style

Cape has come to mean just about any one-story house with a gabled roof. But the original Cape Cod houses, which began appearing in New England in the late 17th century, speak a common architectural language derived from climate, culture, and available materials.

Small in scale and low to the ground, they had short first stories with walls just a bit taller than door and window openings. Roofs pitched at about 8-in-12 kept the weather out and created a usable attic space. Roof overhangs were small, giving strong coastal winds less purchase to do damage. A large chimney centered over the front door served the multiple fireplaces needed for heating and cooking. Roofs and walls were generally covered in wooden shingles, although clapboards were often seen on inland houses and on the front walls of swankier coastal versions. Shutters were common, for obvious practical reasons, but were not the rule.

The label "Cape Cod house" is attributed to the Rev. Timothy Dwight, who traveled through New England around 1800. Other early fans included Henry David Thoreau, who was charmed by the wide variety of windows in their gable ends.

Old Capes have three basic front elevations (see drawings right). The original Crane house was a half-house. All three basic types were designed so that they could be easily and frequently expanded. Half-houses were sometimes converted to one of the larger versions, but the most common addition to all three types was an ell at the back or side. One ell usually led to another, a model I used for the Crane house. —J. B. L.

Half-Cape

Three-quarter Cape

Full Cape

A long hall connecting the kitchen–living room area with the first-floor bedrooms also does double duty. I borrowed 3 ft. behind the bath on one side of the hall for a built-in seat and 4 ft. from the deck on the other side of the hall to create a small sitting room (see photo top left facing page). At about 12 ft. by 10 ft. and ringed by built-in bookcases, the sitting area is just large enough for traffic to pass through without disturbing anyone who happens to be sitting there.

In the end, the rigorously edited final design managed to fit four bedrooms, a living room, a dining room, a kitchen, and three and a half baths into 2,100 sq. ft. At

A vaulted ceiling makes the room look big. The master bedroom, a less-than-palatial 13 ft. by 16 ft., looks roomy and spacious thanks to a vaulted ceiling. Photo taken at G on floor plan.

Changes in the ceiling plane throughout the house are an important part of the design strategy.

Ceiling height goes up for more light. A new living room got a vaulted ceiling with big skylights. The built-in cabinet over the fireplace opens for a television. Photo taken at D on floor plan.

this drastically reduced size, the original 24-ft. by 28-ft. half-house reemerged as the dominant element. Although all the additions were built at once, the finished house is in the tradition of an old Cape that had grown over the years: an informal grouping of similar gabled boxes tied together by a common vocabulary of details (see photo bottom p. 78).

High Ceilings and Lots of Light Help Small Spaces to Seem Voluminous

Changes in the ceiling plane throughout the house are an important part of the design strategy. While the kitchen and dining room have the 7-ft. 3-in. ceiling of the original house, new rooms are made larger

Although all the additions were built at once, the finished house is in the tradition of an old Cape that had grown over the years: an informal grouping of similar gabled boxes tied together by a common vocabulary of details.

in volume with vaulted ceilings. The ceiling of the mid-hall sitting room, for example, keeps the space from being cramped even when several people are using it. Small bedrooms on the first floor, with windows only on the north side, have operable skylights on the south side of the roof to add light, space, and cross ventilation. A vaulted ceiling opens up the master bedroom (see photo p. 80) and two windows high on the west gable wall bring in lots of extra light.

Nowhere does this high-ceilinged approach work better than in the living room (see photo p. 81). The two large skylights and seven tall windows make the room act like a lantern, casting daylight back across the kitchen and dining room. A bright, vaulted ceiling distinguishes the room from the adjacent kitchen, but the large opening between the two rooms ties the spaces together. Living-room windowsills are 18 in. above the floor—high enough to create a sense of boundary between the indoors and out, yet low enough for a seated person to see the ground outside. The biggest of the windows in this room, a triple unit on the west wall, frames a view across the deck and back into the house through the sitting room. Such long views are another device that helps to create a feeling of space greater than the size of the house.

Details and Casework Blend Spaces into a Pleasing Whole

One of the basic goals of the project was to make the new house more than just a beach house. So to bring a kind of order to the project, I used more formal, traditionally inspired details in conjunction with its casual floor plan. Each of the public rooms is anchored by substantial casework and outlined with moldings. A large cabinet

over the prefabricated, zero-clearance fireplace in the living room, which doubles as a media center, is overscaled but very much in the New England tradition. Its built-up cornice, simple flat paneling, and wide mantel moldings are repeated across the kitchen at the dining-room fireplace. And the 4¼-in. crown molding at the fireplace cornice continues throughout the flat-ceilinged dining room and kitchen, with mitered returns punctuating each window, door, and room opening. Window and door casings are simple, elegant, and big.

Although I now spend more of my time designing houses, I started my building career as a carpenter, something I've yet to give up. And when it came time to trim out the Cranes' house, I didn't bother looking for a subcontractor. I set up my portable shop in the three-bay garage and did all the trim and cabinet work on site. The prominent casework, millwork, and trim bring focus to the plan and a good deal of pleasure to the house. They also were fun to build, especially because they went together quickly and produced distinctive results at ordinary prices.

I bought the big poplar crown, but I made all the other molding on my table saw and two table-mounted routers. Except for the poplar bullnose edge on the mantel and window stool, all the trim is made from medium-density fiberboard (MDF) (see sidebar facing page). I made casings that required only butt joints and used material of different thicknesses to make it a little more interesting to look at. The casings are ¾ in. by 3½ in. with a ⁵⁄₁₆-in. bead on each side. Head casings are a full 1 in. by 3¾ in. with a ⅜-in. bead on the bottom edge. On top of each head casing is a ¾-in. by 1⅜-in. cap with a ⁵⁄₁₆-in. fillet and ⅜-in. radius quarter-round profile.

Exterior trim is simple, too. A detail my father used on the original house (and that

Lightweight MDF a Perfect Choice for Interior Trim

Most of the trim is MDF. With the exception of the poplar crown and some bullnose edging, all the trim in the house is made from lightweight MDF.

For me, medium-density fiberboard (MDF) was a natural choice for interior trim. Because MDF mills so easily, I could rout the returns right on the ends of the head casings and caps. Because it is dimensionally stable, the butt joints stay closed without biscuits. Because it has no grain, nailing never splits it. For about $25, a 4x8 sheet of MDF will yield 104 lin. ft. of 3½-in.-wide stock. After it is ripped to width, it routs easily, and most profiles can be cut in one pass. I milled all the trim for the house in a couple of dusty, monotonous afternoons.

Fine, copious dust is one of the two drawbacks to MDF. The other is weight. A sheet of ¾-in. material weighs about 100 lb., quite a bit when you are working alone. To spare my back, I use a product called TruPAN Ultralight, which is made in Chile from radiata pine trees. I bought the Ultralight from Connecticut Plywood; the importer is Tumac Lumber Corporation in Everett, Washington (see Sources for both). It's about 10% more expensive than regular MDF, but it weighs only about 60 lb. a sheet. Although it doesn't mill quite as cleanly as the denser stuff, I can move the sheets around much more easily. To me, the extra sanding it needs (a light pass with 150-grit paper) is well worth the trade in heavy lifting. After sanding, a coat of oil-based primer prepares it for a smooth finish. Latex primer makes the surface rough. —J. B. L.

I repeated) are the chimneys painted with a broad Tory stripe (in Revolutionary times, a detail intended to identify those loyal to the Crown). The wood chimneys disguise either a block chimney or metal flues. They are made from plywood and 2x material—built, fiberglassed, and painted on the ground and then hoisted into place after the roof is complete. I added a texture material to the resin to make the wood look like stucco. A prefabricated flue cap and a copper shroud at the top keep the wood parts safely away from the heat.

Joseph B. Lanza designs and builds in Cambridge and Chatham, Massachusetts.

SOURCES

Connecticut Plywood (for TruPAN Ultralight)
West Hartford, CT
(860) 953-0060

Tumac Lumber Corporation (Importer)
Everett, WA
www.tumac.com
(800) 982-9238

Remaking an Old Adobe in the Territorial Style

T HE OLD ADOBE THAT JEAN AND GARY CLINTON BOUGHT had been built by one of the earliest settlers to the Valdez Valley just outside Taos, New Mexico. Added onto many times since its construction in the late 1800s, the house was still too small and in real need of a rehab (see photo bottom facing page). The Clintons wanted to put on a sizable addition and restore the house so that it would reflect the valley's Spanish farming and ranching character.

To anyone thinking adobe houses are square, flat-roof structures, the territorial style of northern New Mexico will come as a surprise. Appearing in the mid-1800s as American influence increased in the rural Southwest, territorial-style houses had traditional adobe walls. But roofs were gabled, metal clad, and steeply pitched. Floor plans were long and narrow, and gabled

LEFT Not all adobes have flat roofs. A pitched roof and gabled dormers are typical details of a New Mexican territorial-style adobe. Photo taken at B on floor plan.

ABOVE A new wing added living space and created a sheltered courtyard at the rear of the house. Photo taken at A on floor plan.

LEFT New life for an aging adobe. A 100-year-old adobe outside Taos, New Mexico, was ripe for an overhaul when it changed hands. Photo taken at B on floor plan.

To anyone thinking adobe houses are square, flat-roof structures, the territorial style of northern New Mexico will come as a surprise.

dormers were common. The design for the addition and renovation—worked out among me, Jean Clinton (who has a background in design), Angela Matzelli of House Floor Plans in Taos, and Donna LeFurgey of Taos Drafting and Design—leaned on this regional style.

With walls of traditional adobe brick, the 1,540-sq.-ft. addition sits at right angles to the original house and creates a sheltered courtyard area at the rear of the house (see photo top p. 85). New living spaces include a large master bedroom and family room, each with high ceilings lit by gabled dormers (see floor plans below). Trim and other exterior detailing, plus a new coat of cement stucco on old and new parts of the expanded house, help to unify the design. The Rio Hondo, a small river running below the house, and the beautiful Sangre de Cristo mountains up the valley to the east are features that needed no improvement.

A Change in Grade becomes the First Problem to Solve

Because of a sloping site, the addition had to sit above the existing house by about 4 ft. When finished, the addition was linked to the existing house by a curved staircase (see photo top facing page). Atop a 24-in.-wide concrete footing, the addition's stemwall of 14-in. block went up between three and nine blocks. Once the footing and stemwalls were done, we poured a concrete floor over compacted earth and a layer of crushed stone. Polystyrene insulation 1 in. thick provides a thermal break in the foundation.

Adobe bricks that form the outside walls are 10 in. by 14 in. by 4 in., laid in cement mortar and capped with a steel-reinforced concrete bond beam. At the top of the adobe wall, two 2x plates anchored to the bond beam tie the roof system to the walls.

GOING UPHILL FOR NEW LIVING SPACE

A sizable addition includes a new entry, a master-bedroom suite, and a family room. Connecting old and new portions of the house is a curved staircase that overcomes a 4-ft. difference in grade on the site.

SPECS

BEDROOMS: 4

BATHROOMS: 3½

SIZE: 4,350 sq. ft.

COST: n/a

COMPLETED: 1996

LOCATION: Taos, New Mexico

BUILDER: Ken Wolosin

North

0 2 4 8 ft.

Photos taken at lettered positions.

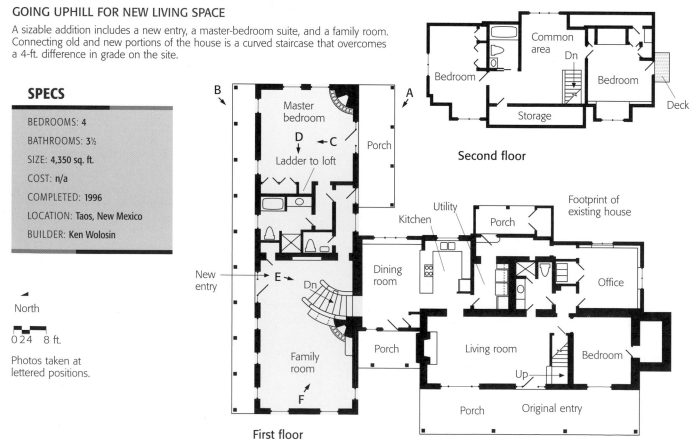

Second floor

First floor

LEFT **Stair links addition to existing house.** A gently curved set of stairs overcomes a grade change to provide access to the new family room. Photo right taken at E on floor plan.

BELOW **A private master-bedroom suite.** At the far end of the new wing, the master bedroom is generously supplied with natural light. Photo taken at C on floor plan.

The 5-ft. kneewall above the bond beam is framed conventionally with 2x10s.

Although adobe has incredible mass, its insulating value is low. So the exterior walls include 2 in. of rigid, foil-faced insulation. In keeping with the architectural style, wood trim is set around the windows. We used Pella Architect Series double-hung windows with divided lites and insulating glass (see Sources p. 89).

The Clintons wanted cathedral ceilings in both the master bedroom (see photo right) and the family room. Across the 22-ft. width of these rooms, we ran large peeled logs called vigas, a traditional element of adobe houses. Set on 4-ft. centers, these structural vigas leave the high ceiling and dormer windows visible from below (see photo p. 88) while serving as collar ties for the roof. The 2x12 roof rafters are covered on the inside with beaded 1x6 tongue-and-

groove boards. The beading, a territorial motif, also is used on door jambs and doors.

With high ceilings at both ends of the addition, the center of the new structure, consisting of a hallway, half-bath, and master bath, has flat ceilings. Above is a meditation loft reached via a kiva ladder on one wall of the bedroom (see photo p. 89).

Reworking the Existing House Helped to Blend the Two Structures

Remodeling on the house was extensive: Vigas were added to one room; an existing bath was gutted, the ceiling raised and completely redone; new heating, plumbing, and electrical systems were installed; new doors and some new windows were added; the old family room got a new oak plank floor; and baseboards and mud plaster were added.

Outside, we cut back the roof overhangs so that they would more closely match those on the addition. We removed the wide chalet-style fascia boards and added narrower trim on all the doors and windows. In the territorial style, head casings have a small peak at the center of the window. And we redid the roof with the same galvanized steel that we used on the addi-

Vigas are functional and add interest. Not only are they collar ties for the roof, but these Vigas also highlight the cathedral ceilings and unify both the home's existing and new interiors. Photo taken at F on floor plan.

A ladder leads to a small loft above the master bathroom. Photo taken at D on floor plan.

tion. Once the exterior walls of the whole project were plastered over with cement stucco, the parts blended.

Detailing Inside and Out Is More Than an Afterthought

Many elements combine to give the house the feel of a traditional Spanish territorial home: tin light fixtures and mirror frames, custom doors, forged door hardware, kiva fireplaces, and thick exterior adobe walls. Mud plaster used to finish interior walls of the addition is especially important. It is a traditional mud and straw mix that is applied over a cement scratch coat.

Before mud plaster was applied, we installed door casings and baseboard so that the finished plaster walls would be nearly flush with the trim. In the bathrooms, wall tiles and cabinets were installed, followed by plaster. Trim work in the house was painted with a white enamel.

To me, among the more interesting features of the project are the tapered posts on the porches, or portals as they are called around here. The design was inspired by the posts of an old Spanish adobe in Jacona, New Mexico, which at one time belonged to the artist Cady Wells. Cabinet and door builder Ed Paul made the posts by wrapping ¾-in. exterior-grade plywood with 1-in. pine protected by West System epoxy from Gougeon Brothers Inc. (see Sources). Concealed rabbets at the bottom of the posts accommodate galvanized brackets that are pinned to porch floors with ⅝-in. threaded rod.

Ken Wolosin has been a builder in Taos, New Mexico, for more than 25 years.

SOURCES

Pella Corp.
Pella, IA
(800) 847-3552
www.pella.com

Gougeon Brothers Inc.
Bay City, MI
(989) 684-6881
www.gougeon.com

A Wish for a Kitchen and a Bath

I T WAS SUPPOSED TO BE A PORCH. WHEN SALLY MANESIS FIRST called me, she was looking for an architect to design a modest screened-porch addition to fit her turn-of-the-century Victorian home. As it was first conceived, the project didn't exactly thrill me, but I didn't hesitate to meet with her. I make it a point to meet with every

Octagonal bump-out with hip roof did the trick.
Grafted onto the side of the house, the expanded
kitchen and master bath look as if they were always
there. The inset photo shows the house before the
bump-out and porch were added. Photos taken at A on
floor plan.

A MODEST TWO-STORY ADDITION REVITALIZES THE MOST IMPORTANT ROOMS IN THE HOUSE

Although its footprint occupies only 180 sq. ft., this octagonal addition includes a versatile eat-in kitchen area on the first floor and a sun-filled master bathroom on the second floor.

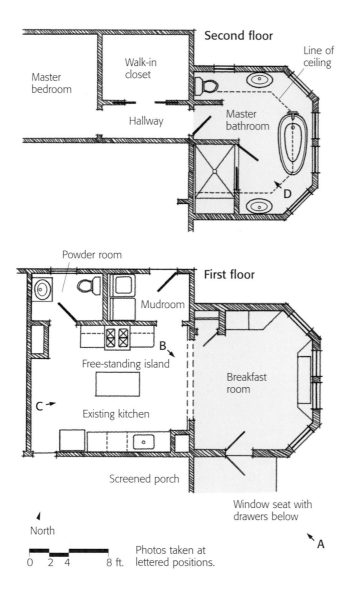

Second floor

Line of ceiling

Master bedroom

Walk-in closet

Hallway

Master bathroom

D

Powder room

First floor

Mudroom

Free-standing island

B

Breakfast room

C

Existing kitchen

Screened porch

Window seat with drawers below

North

0 2 4   8 ft.

Photos taken at lettered positions.

A

Porch Was the Tip of the Iceberg

Although Sally and Andy's new screened porch was at the top of their project list, bathroom and kitchen improvements weren't far behind. The kitchen cabinets, countertops, and appliances had been tastefully upgraded by the previous homeowner, so the basic work triangle was fine. But the kitchen did not have enough storage space or enough counter space for the large-scale cooking, baking, and entertaining that Sally enjoys. Sally's wish list for the house also included an eat-in kitchen—a space for informal entertaining and for displaying her collection of antiques from her birthplace in Lancaster, Pennsylvania.

Unlike the kitchen, the existing master bathroom had no redeeming qualities whatsoever. Definitely not original to the house, it was most likely added during the 1950s as a second bathroom. This tiny space, with a shower stall that could have easily been mistaken for a telephone booth, could never be a master bathroom by today's standards. Fortunately, the bathroom was located directly above the kitchen, making it relatively simple for us to tackle both improvements at the same time.

Half-Octagon and Wraparound Porch Make Addition Feel at Home

Ultimately, Sally and Andy gave me the go-ahead to expand the house for the sake of the kitchen and master bathroom, as long as I could find a place to put the porch. My solution was to bump out the east wall of the kitchen and create a two-story addition that would include a breakfast nook on the first floor and a master bathroom on the second (see floor plans left).

prospective client, partly out of courtesy, partly because I know that construction projects tend to evolve.

Old Victorian houses have great charm and character, but they were not built to accommodate today's lifestyles. Something always needs to be upgraded or replaced. I was not surprised to discover that Sally and her husband Andy's wish list included more than just a screened porch.

Great Idea: New Cabinet from Barn Salvage

While browsing in a shop near Philadelphia, Sally found a great bathroom-storage solution. Sally's four-pane medicine cabinet looks like an antique because it is, sort of. For more than ten years, furnituremaker Bryce Ritter has been making period designs (and some original pieces) using antique hardware, lumber, and millwork, all of which are salvaged from area barns. Ritter's creations can be purchased through Pennsylvania Traditions (see Sources).—**L. D.**

Recycled antiques. An intact barn window, complete with wavy glass, completes the illusion that the medicine cabinet is a rustic heirloom. Photo by Tom O'Brien

---

The addition's hip roof and octagonal footprint were designed to be compatible with the smaller bay on the front of the house (see photo pp. 90–91). A new gable roof behind the addition's hip roof balances the rooflines and replaces an existing dormer.

I created a wraparound porch by extending the original front porch around the side of the house, terminating at the addition. The new side porch not only helped the addition fit in with the original house but also proved to be the perfect place to include the long-sought-after screened porch.

During the construction process, I was able to realize one of my own wish-list items: The clients—with some prompting from contractors Jim Goebel and Peter Marschalk—opted to remove the inappropriate vinyl siding from the existing portion of the house and restore the original shingle and clapboard siding and the detailed wood trim. These same wood features were replicated on the addition, ensuring that it would further integrate with the original house.

The addition is supported by a concrete foundation that is covered by stone veneer matching the house's stone foundation. We got all the stone that we needed as a byproduct of excavation.

Landlocked Kitchen Breaks Free

The existing kitchen was essentially left intact. With its freestanding island, granite countertops and mostly new appliances, the galley-style kitchen more than adequately serves its basic functions. At Sally's request, the electric range was replaced with a more serious piece of cooking equipment: a state-of-the-art dual-fuel range by Heartland (see Sources) that only looks as though it belongs in the 1930s (see photo top p. 94). The other improvements made to the existing kitchen were minor. These included new cabinet hardware, new light fixtures, refinished hardwood floors, and a fresh coat of paint.

With its free-standing island, granite countertops, and mostly new appliances, the galley-style kitchen more than adequately serves its basic functions.

Kitchen gets a new range and a lot more space. A state-of-the-art range (hidden beneath a 1930s exterior) improves a perfectly functional, if small, galley kitchen. The addition (seen in the background) adds storage space and a cooking station as well as a dining area. Photo taken at C on floor plan.

A country kitchen in the city. Farmhouse table, Amish rugs, and an antique chicken incubator, which now serves as a wine rack, fit perfectly within the informal octagonal space. The sought-after screened porch is the door on the right. Photo taken at B on floor plan.

The octagonal shape of the kitchen addition allows for plenty of windows with varied views and creates a space for Sally's farmhouse table and Amish rug (see photo left). The south wall was perfect for Sally's wine rack, which is an antique chicken incubator, and the door to the screened porch. The addition also includes a coat closet and a window seat with drawers. A workstation on the north wall of the addition includes a built-in oven, a microwave and a prep area.

A Master Bath at Last

Whereas the kitchen was merely improved, the original master bathroom was gutted and used to expand an existing walk-in closet and to create a hallway to the new bathroom.

As with the kitchen below, the octagonal addition provides a spacious, sun-filled setting for the new master bathroom. Its prominent features include oversize double-

As with the kitchen below, the octagonal addition provides a spacious, sun-filled setting for the new master bathroom.

Soak your cares away. An oversize clawfoot tub with faucets on the side, lumbar support in the back, and rubber duckies in the front is the perfect antidote to a long day in the city. His and hers pedestal sinks are at opposite ends of the room. Photo taken at D on floor plan.

hung windows, painted beadboard wainscoting and a lofty vaulted ceiling (see photo above).

The focal point of the space is the oversize clawfoot soaking tub. His and hers pedestal sinks are positioned on each side of the tub (see Sources for both). The huge walk-in shower includes frameless glass doors, a bench seat, and enough space to hose down a rugby team.

Excluding the shower, all the fixtures are reproduction pieces. The rustic barn-board

medicine cabinet is a new piece of furniture made of antique parts (see sidebar p. 93). The Victorian trim elements (upstairs and down) are almost identical to those throughout the main house, but they were actually built up using stock moldings.

Louis A. DiBerardino is an architect and owner of Studio DiBerardino in New Canaan, Connecticut. The firm's projects often involve improvements to historic homes.

SOURCES

Heartland Legacy (Range)
(800) 361-1517
www.heartlandapp.com

Porcher Epoque (Clawfoot tub)
(800) 359-3261
porcher-us.com

Lebijou Vitra (Pedestal sinks)
(800) 635-6879
lebijoucollection.com

Kallista (Tub and sink faucets)
(888) 452-5547
www.kallista.com

Pennsylvania Traditions (Bathroom cabinet)
Skippack, PA
(610) 584-8218

A New Approach to the Kitchen

I t was a common situation: The house, a Dutch colonial, had a charming front door that nobody used. The back door, as in most homes, was the primary entrance but lacked in both function and form. It had an uncovered stoop and opened directly into the kitchen. It had no sheltering overhang, no place to put a bag of groceries while the homeowners were fumbling for keys, none of the grace of the front door. Fittingly, when you opened the back door, it blocked the refrigerator.

Breaking down the kitchen walls. In the original design, a wall separated the kitchen and dining room in this Dutch colonial. In the new layout, the rooms are open, and the kitchen is entered through a mudroom (see photo above).

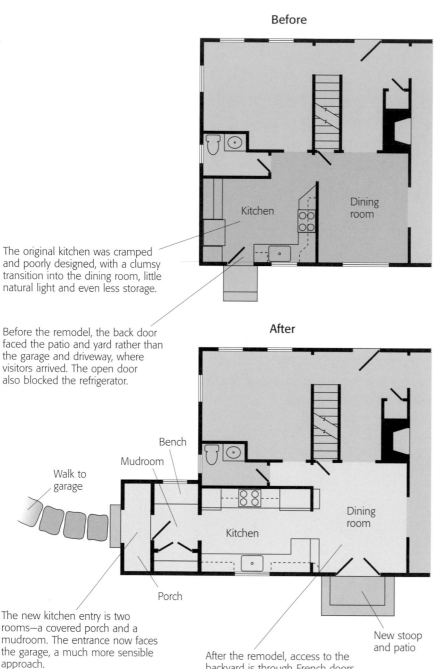

Before

Kitchen

Dining room

The original kitchen was cramped and poorly designed, with a clumsy transition into the dining room, little natural light and even less storage.

Before the remodel, the back door faced the patio and yard rather than the garage and driveway, where visitors arrived. The open door also blocked the refrigerator.

After

Walk to garage

Mudroom

Bench

Kitchen

Dining room

Porch

The new kitchen entry is two rooms—a covered porch and a mudroom. The entrance now faces the garage, a much more sensible approach.

After the remodel, access to the backyard is through French doors in the dining room. The wall separating kitchen and dining room has been removed.

New stoop and patio

The kitchen was equally flawed. It was cramped, with little natural light, and the cabinet and countertop spaces were insufficient by today's standards. There was no connection to the dining room; the two areas were separate, enclosed rooms joined by a clumsy hallway (see floor plans left).

This layout, typical of the builder homes of the 1970s, is a common problem. Today, we spend more time in the kitchen, and our entertaining has become more casual. Kitchen and dining spaces are less separate and more utilitarian. Formal dining rooms—separate, enclosed spaces—are out of fashion.

So the client who came to me with pictures of this Dutch colonial wanted to solve two problems: He wanted a new back door and a new kitchen within the existing space. I proposed a different approach: a sensible entrance consisting of a porch and mudroom, and a well-lit, inviting kitchen that opened to the dining room (see photo p. 96).

Elements of a Good Entrance

What makes a good entrance to a home? Shelter from wind and rain. A seat for taking off boots. A table to empty pockets or to rest a package. A corner to drop an umbrella. Hooks for hanging coats. A closet for storing gear. Elbow room to accomplish all these tasks comfortably. These are the ingredients of the classic New England mudroom. A mudroom is a catchall entry. It's a place to put on outerwear and to store snowshoes and tennis rackets. It serves as the home's filter, capturing dirt, mud, salt, and sand.

Where should the mudroom go? In this case, the original location of the back door—facing the backyard—did not make sense. People come and go from the driveway and garage, so it seemed clear that the

The covered entry provides shelter while the homeowners fumble for keys or shake an umbrella.

Because mudrooms are a transition between outdoors and inside, they can run the gamut from open porch to fully heated and insulated enclosure.

mudroom and kitchen entrance should be near the garage.

Because mudrooms are a transition between outdoors and inside, they can run the gamut from open porch to fully heated and insulated enclosure. We did both with this space. Part of it is an enclosed outdoor porch so that those entering can enjoy cover while unlocking and opening the door (see photo above). We also included room for a small table, snow shovels, and other such gear. Inside the door, there is a roomy space with shelves, hooks, a closet, and a window seat with a hinged lid for recyclables (see photo p. 97).

To dress up the entry, we added a cornice return topped with copper flashing (see bottom photo p. 100). On the roof, the bottom course of asphalt shingles was built up with two layers of cedar. We used an elliptical arch over the entrance and added a small window for light.

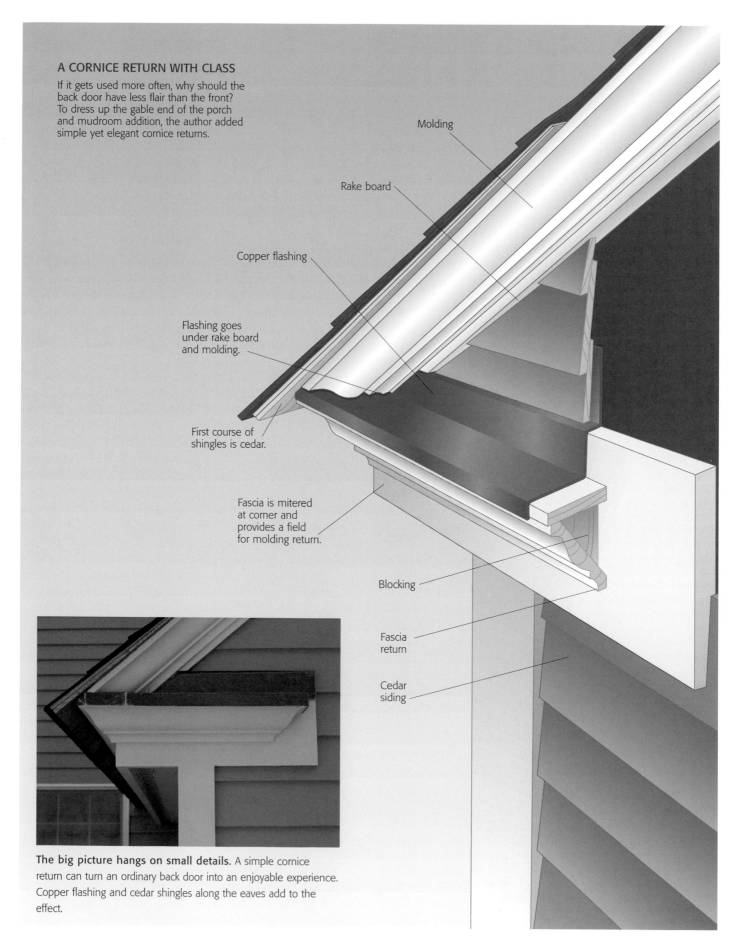

A CORNICE RETURN WITH CLASS

If it gets used more often, why should the back door have less flair than the front? To dress up the gable end of the porch and mudroom addition, the author added simple yet elegant cornice returns.

Molding

Rake board

Copper flashing

Flashing goes under rake board and molding.

First course of shingles is cedar.

Fascia is mitered at corner and provides a field for molding return.

Blocking

Fascia return

Cedar siding

The big picture hangs on small details. A simple cornice return can turn an ordinary back door into an enjoyable experience. Copper flashing and cedar shingles along the eaves add to the effect.

Smooth transition from garage to kitchen. The new back door and mudroom face the garage, a good place considering that the owners arrive and depart by car.

A Kitchen Must Flow

We changed the kitchen layout completely. First, we closed off the old back door and turned that section of wall into counter space for the kitchen. Then we removed the partition between the kitchen and the dining room, replacing it with a 42-in. high peninsula and a built-in china cabinet. These elements opened the kitchen to the dining room while maintaining a transition from space to space. The dining room remained the same, except that we replaced a large window with French doors that open onto a brick patio. This entry replaced the old back door for access to the backyard.

The redesigned kitchen has a galley layout, its narrower workspace flanked by cabinets and appliances. Although the size of the kitchen remains the same, open floor space actually has been reduced by about one-quarter. That change allowed the addition of storage and counter space the owners really needed.

Intimacy Makes Good Architecture

I am a furniture maker in addition to being an architect, and I built the kitchen cabinets, the closet and entrance doors, and the cherry trim. The builder, Pete Mitchell, did the sitework and subcontracting. We both helped on the cabinet installation and finishing.

For me, working with the materials kept my designs grounded in reality. It also allowed me to change details in midconstruction. Certain details, no matter how good they look on a blueprint, are not appropriate when the time comes to build them.

Andrew Peklo III, owner of Pelko Design and Joinery, P.C., is an architect and joiner in Woodbury, Connecticut. Abby Greenwald, a freelance writer, helped with this chapter.

A New Master Suite

I designed my first home 10 years ago on a small lot that everyone else had understandably passed over. Recently married and on a tight budget, we thought the third of an acre on the side of a wooded ravine offered great possibilities. And there we built a 2,700-sq.-ft. home that reflected our love of nature and Japanese gardens and our appreciation for the Arts and Crafts style American bungalow. We might have found the house roomy enough indefinitely. Two children later, though, it seemed too small.

In the original plans, we had a living room but no family room. I had naively believed we wouldn't need one. But as our children got older, they gradually took over our first-floor master bedroom. Eventually, we decided to convert this room to its de facto use and add a master suite. The result was a 1,000-sq.-ft. adult retreat (see photo p. 104) that gave us all more breathing room. Although attached to the original house (access is through the old master bedroom), the new suite seems more like a secluded cottage.

A refuge from busy family life. A new master-bedroom addition is a private hideaway with a wall of windows facing a small Japanese-style garden and a wooded ravine. Photo taken at A on floor plan.

A Hilly Site Leads to a Two-Tier Design for the Addition

Our neighborhood, a few miles from downtown Knoxville, Tennessee, had been built in the 1930s. Our small lot had been overlooked because it was in a wooded ravine that made building a challenge. The site's topography put the first floor roughly one story above the level of the attached two-car garage (see floor plans p. 106). A stairway connects the garage with the first floor of the house.

When it came time to design the addition, the site worked to our advantage. It made sense to push the new master-bedroom wing off the back of the garage and into the backyard. On the lower level, the addition offered room for a guest bedroom and full bath connected to the garage by a covered patio. The patio not only provides

additional outdoor living space but also offers valuable covered storage for bikes and garden equipment.

Upstairs, we broke through the back of the walk-in closet in the old master bedroom to create an entry for our new suite. A skylight provides daylight to this efficient yet windowless entry/closet corridor. Because the elevation of the new suite's main floor is the same as the first floor in the existing house, the connection is seamless.

A Plan That Makes the Most of Its Natural Context

Above all else, I wanted the addition to take advantage of a site dotted with oak trees more than a century old. Large window walls were a first step toward making an outdoor connection. A north-facing, 12-ft.-high wall of windows gives us unobstructed

This master suite is an adult retreat, complete with sleeping and sitting areas, master bathroom with separate bathing area, and access to the deck and garden.

Our appreciation for the colors and textures of traditional Arts and Crafts buildings showed in the clear redwood that surrounds the tub and in the custom-made tile.

LEFT A big tub with a view. A 6-ft. by 6-ft. tub surrounded by awning-style windows makes an inviting corner in the master bathroom. Photo taken at D on floor plan.

views of our neighbor's wooded ravine. I like to think of the view as a "borrowed landscape," and it makes a great backdrop for our small circular garden. Walls on the east and west sides of the small sitting alcove also are windowed.

I decided to zone the new master bathroom, separating the shower/tub area from a private toilet alcove that is easily accessible from the sleeping area. A bathing area—with a large, tiled shower and a 6-ft. by 6-ft. tub— is lighted by two large skylights as well as a band of awning-style windows that offer ventilation and views while still managing to provide some privacy (see top photo).

LEFT Handmade tiles and redwood add color and texture to the room. Photo taken at E on floor plan.

An addition built on the north side of the existing house includes a master bedroom, sitting alcove, and bath on the first-floor level, and a guest bedroom and bath on the lower level. Access to the new master suite is through what had been a walk-in closet.

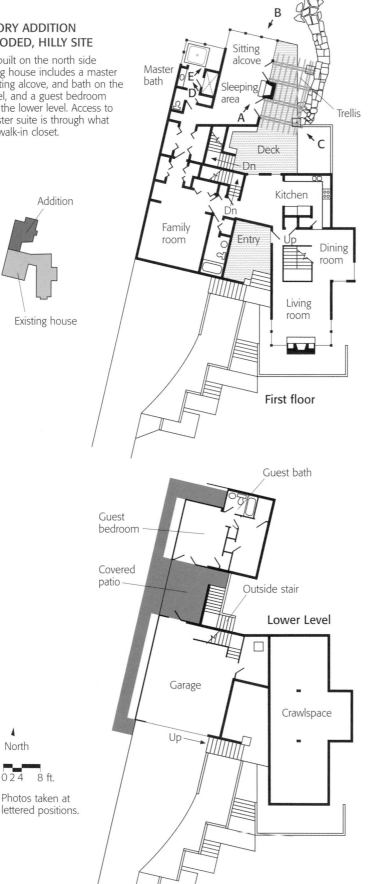

Addition

Existing house

First floor

Guest bath

Guest bedroom

Covered patio

Outside stair

Lower Level

Garage

Crawlspace

Up

North

0 2 4   8 ft.

Photos taken at lettered positions.

Our appreciation for the colors and textures of traditional Arts and Crafts buildings showed in the clear redwood that surrounds the tub and in the custom-made tile (see top photo p. 105). The redwood makes a soft, warm tub platform, and it is naturally resistant to water damage.

As in the main house, walls in the addition are lower than usual. I think they make the scale of the building more intimate.

A Trellis and Deck Connect the Addition to Outside Spaces

Unlike Western design, which treats a garden as an object separate from a walled house, the Japanese approach is to make the garden an extension of the house. There is no clear demarcation between inside and out. Instead, a third type of space is created, an intermediate space that connects the inside with the outside. Japanese builders make extensive use of verandas, shoji doors,

A stone lantern honors Japanese design influence. Photo taken at C on floor plan.

Cedar trellis connects home to landscape. A low roof over French doors and a cedar trellis are a transition between house and garden.

and long, low rooflines. So important is this idea that the Japanese even have a special word for it: *nokishita,* which means space under the eaves.

I tried to put these principles to work in designing our new master suite (see photo p. 104). Large windows are a start. Two large French doors flanking the bedroom's fireplace lead to the deck outside. They are sheltered by the addition's shed-dormer roof, cantilevered 5 ft. from the sidewall.

Tucked under this protective overhang is a cedar trellis that extends over the deck and into the landscape (see photo above). The effect is of an intimate exterior room. I think the layering of roof eaves and trellis blurs the understanding of what's inside and what isn't.

Dr. Scott A. Kinzy is an architect and a professor of architecture at the University of Tennessee in Knoxville.

Remodeling a Cape for a Shrinking Family

W E'VE BEEN LUCKY. MY WIFE JANICE AND I HAVE LIVED IN a great 1,160-sq.-ft. Cape for 25 years, and we plan to stay. Grand 50-year-old elms that escaped the elm blight line our street. (Edmonton boasts the continent's largest stand of uninfected American elms, with about 60,000 elms throughout the city.) Although we raised our children here, we didn't have to move to a bigger house. Their departure to start their own lives has been a major turning point for all of us. So three years ago, with one child out of the house and the other growing fast, we decided that renovating would help us to make the transition back to life as a family of two. We added only 240 sq. ft. inside but made much better use of the space. The whole renovation took more than two years' worth of weekends and holidays, and cost about $26,000 (US).

Two dormers added only 240 sq. ft. inside, but the dormers and a veranda transformed the exterior. Photos taken at A on floor plan.

Better Use of Space Makes the House Serve the Owners

We moved our bedroom upstairs, turning our old bedroom into a bigger dining room. A hallway separated the small existing living and dining rooms, but our love of cook-ing and entertaining called for larger rooms that flowed together. Putting the dining and living rooms together and removing the wall between them solved the problem. The small existing dining room became a study.

The two bedrooms upstairs huddled under the roofline, and we expanded them

With the dormers and veranda framed and the house reroofed, the author could complete interior construction on weekends and holidays. Photo facing page taken at C on floor plan.

by adding a shed dormer in back (see photos facing page and right) to give us a little extra headroom, floor space, and built-in storage. A small gable dormer placed in the front let us turn two back-to-back closets into a second bathroom.

We bumped out the confined front entry just enough to make it into a small foyer with better storage for coats and hats. While the open plan and dormers gave us more space on the inside, the veranda changed the exterior of our little Cape, extending the living space outward so that we could enjoy

Remodeling a Cape for a Shrinking Family   **111**

A veranda adds outdoor living space. Balusters from recycled decking and a beveled 2x4 handrail with cove molding gave a custom look without stretching the budget. Photo taken at B on floor plan.

visiting neighbors and friends over tea or a cool beer.

Some of the friends and family we hoped to entertain better turned out to help with the heavy lifting. It took a group effort to frame the veranda and dormers and to reroof the whole house. This early help made the house weathertight and allowed me to work on the interior renovations at my own pace.

Built-Ins and Closets Add Storage in Tight Spaces

In our daughter's bedroom, the extra 6 ft. we gained by adding the shed dormer provided space for a new closet, pullout shoe storage, and a window seat that houses an aromatic cedar-lined chest (see bottom photos facing page). Display shelves and book-shelves on the walls of the window seat

LEFT AND ABOVE **Closets use eaves space.** Hanging bars and shelves cross behind the dresser to join the two wardrobe-fronted closets. Photos taken at D on floor plan.

ABOVE **Use every inch.** Full-extension drawer slides permit shoe storage in the eaves space. Photo taken at E on floor plan.

LEFT **The window seat lifts to reveal a cedar chest,** and built-in shelves use the last few square feet of wall. The closet at right is also new. Photo taken at E on floor plan.

allow her to dream and read there. Although Maggie is close to leaving, I can imagine reading stories to our grandchildren in this cozy little spot.

After living in a small first-floor bedroom for so many years, Janice and I wanted our new bedroom to be a bright space. Instead of using the shed dormer to add closets, we simply put in a few shelves and a big win-dow to read by. I did build two closets and a dresser into the previously unused eaves space along the front of the house, however (see top photos). I had planned to build paint-grade closets out of medium-density fiberboard. However, I lucked into four clear-pine panel doors in a clearance bin, so I purchased a few boards and built the closets and dresser around these doors.

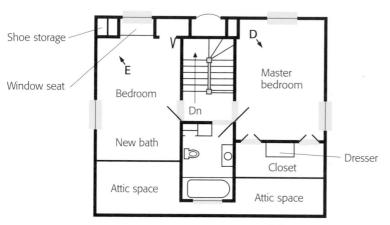

Shoe storage

Window seat

E

Bedroom

New bath

Attic space

V

Dn

Master bedroom

D

Dresser

Closet

Attic space

Second floor after

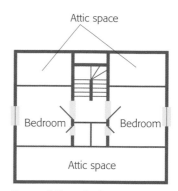

Attic space

Bedroom

Bedroom

Attic space

Second floor before

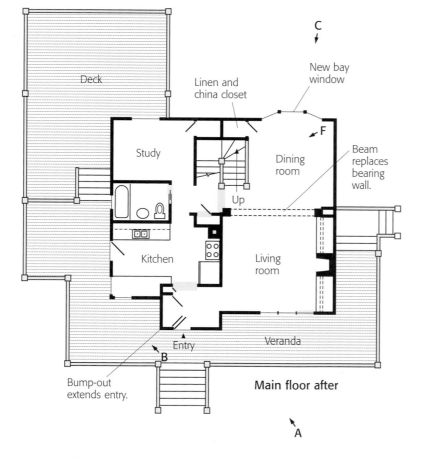

Deck

Study

Linen and china closet

New bay window

C

Dining room

Beam replaces bearing wall.

F

Up

Kitchen

Living room

Entry

Veranda

B

Bump-out extends entry.

Main floor after

A

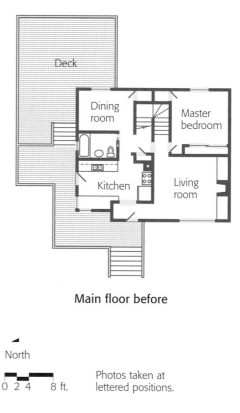

Deck

Dining room

Master bedroom

Kitchen

Living room

Main floor before

North

0 2 4   8 ft.

Photos taken at lettered positions.

Opening the Main Floor Makes Entertaining Easy

To join the new dining room to the living room, I removed the bearing wall that separated them (see photo facing page). A couple of friends came back one day and helped to build temporary support walls on both sides of the bearing wall to shore up the floor above. Then we replaced the wall with a beam made of three 2x12s that rest on new columns. (Directly below in the basement, posts carry the load to concrete

Big Changes on a Budget

Opening up the first floor allowed the master bedroom to move upstairs and yields free passage between the living and dining rooms (photo taken at F on floor plan). Dormers add space, storage, light, and a second bath upstairs. Built-in storage makes all the spaces work harder, including the enlarged foyer. The new veranda affords lots of space for relaxing with friends. The renovation, including new hardwood floors and other incidental changes, cost $26,000 (US).

pads.) We set the top of the beam level with the top of the existing floor joists. That left about 3 in. of the beam projecting below the ceiling. I trimmed this 3-in. beam projection with shallow arches that match those on the veranda. The room feels like one big open space now, with the arch defining the passage from living room to dining room.

Howard Pruden has been involved in construction for 30 years, and a fire inspector in Edmonton, Alberta, Canada for the last 22.

Building a Grand Veranda

OUR CLIENTS' HOME WAS A 19TH-CENTURY THREE-STORY manor near the banks of the Delaware and Raritan canal in central New Jersey. Built for the owner of a nearby rubber factory, the house had triple-wythe exterior brick walls 12 in. thick. The house also boasted a large porch that wrapped around three sides.

In its second century, the house began to show its age and needed significant repairs. In the 1960s, the porch succumbed to rot and decay and was removed. When our clients bought the house, much-needed interior work kept the porch project on the back burner. But images of leisurely summer evenings dining on the porch overlooking the canal with rockers creaking finally inspired them to rebuild the porch.

Old photos tell a porch's tale. These photos saved by the owner were the only record of the porch. The top left photo reveals the graceful arches, but also the flat roof that caused the porch to fail. The top right photo shows the empty girder pockets and the tar line of the old roof.

The porch makes the house. Without the porch, the three-story brick walls made this house look stark and barren. The porch provides a graceful skirt around three sides, turning barren into beautiful (photo facing page).

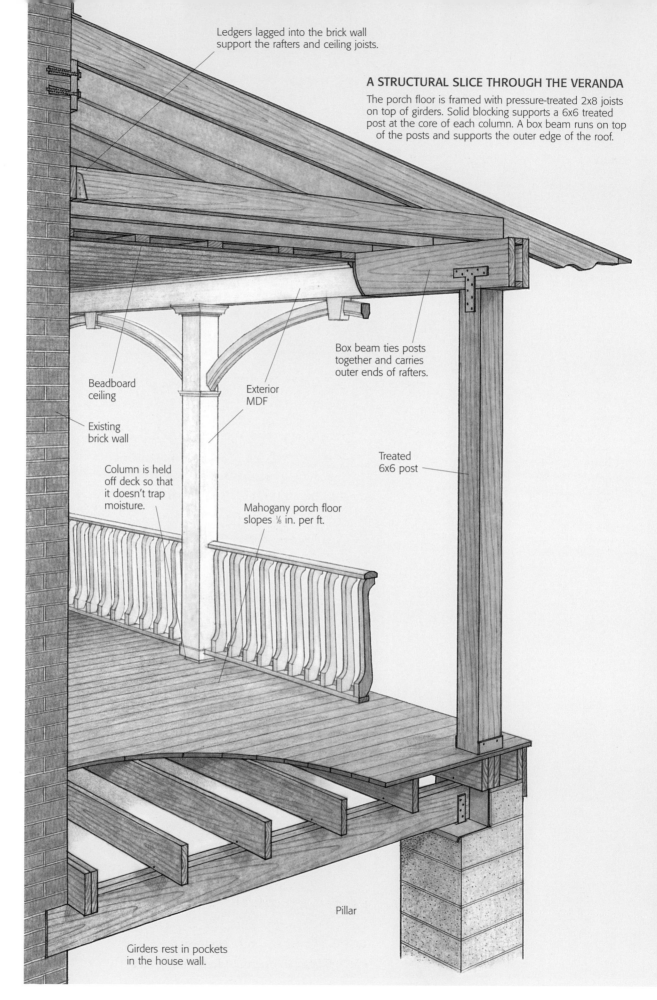

Ledgers lagged into the brick wall
support the rafters and ceiling joists.

A STRUCTURAL SLICE THROUGH THE VERANDA

The porch floor is framed with pressure-treated 2x8 joists
on top of girders. Solid blocking supports a 6x6 treated
post at the core of each column. A box beam runs on top
of the posts and supports the outer edge of the roof.

Box beam ties posts
together and carries
outer ends of rafters.

Beadboard
ceiling

Exterior
MDF

Existing
brick wall

Treated
6x6 post

Column is held
off deck so that
it doesn't trap
moisture.

Mahogany porch floor
slopes ⅛ in. per ft.

Pillar

Girders rest in pockets
in the house wall.

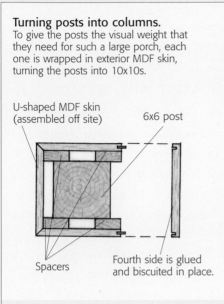

Turning posts into columns.
To give the posts the visual weight that they need for such a large porch, each one is wrapped in exterior MDF skin, turning the posts into 10x10s.

U-shaped MDF skin
(assembled off site)

6x6 post

Spacers

Fourth side is glued
and biscuited in place.

**Chimney-block pillars
hold up the porch.**
The perimeter of the porch is supported on 20 pillars. First, concrete footings are poured. Then 16-in. chimney block is stacked with rebar in the hollow cores. The cores are filled with concrete, and the outside of the block is parged with plaster.

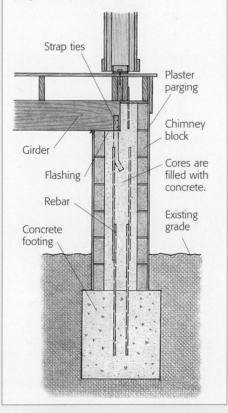

Strap ties

Plaster
parging

Chimney
block

Girder

Flashing

Cores are
filled with
concrete.

Rebar

Existing
grade

Concrete
footing

Rediscovering History

The only record of the original porch was a photograph, faded and dog-eared, that the owners had saved while waiting to rebuild the porch (see left photo p. 116). It was difficult to make out all the original detailing from the photograph, but we could see the flat roof as well as the gentle arches curving sweetly across each bay.

In addition to the photo, we found further clues on the existing exterior walls of the house (see right photo p. 116). The brick was corbeled out to create a drip edge, a detail that gave us the finish-floor height. In addition, vacant girder pockets that once held the porch floor and ceiling joists sat like pockmarks in the brick wall. In the ground, remnants of piers indicated the porch dimensions and the column rhythms, and the long black line that scarred the brick wall beneath the second-floor windows indicated where the porch roof had been sealed to the house with tar.

Porch becomes a Veranda

Armed with these details, our crew here at Princeton Design Guild began to redesign the porch to create a generous, comforting edge to the severe-looking three-story brick wall. More than just a porch, we designed a veranda 8 ft. deep and more than 160 ft. long.

The dining room, living room, family room, and kitchen all open onto the veranda from three sides of the house. This new corridor around the perimeter of the house gives each of these spaces a generous new anteroom to the exterior world, creating new and flexible traffic routes from the outside.

The west side of the porch, which faces the canal, is great for afternoon rocking in the setting sun (see photo p. 121). The southwest corner, where the house jogs back, creates an open outdoor dining area.

The flooring was run perpendicular to the house wall so that any water runs off parallel to the seams, not across them.

A wide, welcoming staircase leads to a formal entrance on the north side, while the east side provides a private play area for the kids.

Piers Made of Chimney Block

We started by digging and pouring 22 footings. Piers that support the porch were built on top of the footings and tied into the rebar cast into each footing (see drawing bottom p. 119). Because the porch columns above the floor had to line up directly over the piers, we had to be extremely precise when locating the piers.

Each pier was built of 16-in. chimney block. As we stacked the block, we tied together four columns of rebar that thread up the center cavity (where the flue would normally have been). We then filled the core of each pier with concrete.

At the top of each pier we added a block to the front face, which allowed us to hide the double 2x10 treated girders that span from each pier to the girder pockets in the brick wall of the house. The pockets, which were made for 19th-century wood, had to be modified and recut to accommodate the wood of the 1990s. We then ran 2x8 treated floor joists parallel to the outside walls on top of the girders, 16 in. o.c. The 1x4 tongue-and-groove mahogany flooring was installed directly on top of the joists.

We gave the girders a slight pitch (⅛ in. per ft.) away from the house wall. Although plans called for a full roof over the porch, I was concerned that blowing rain or snow could end up on the floor. The pitch ensured that any water would drain away safely.

The flooring was run perpendicular to the house wall so that any water runs off parallel to the seams, not across them. We sealed the floor and let it cure for two days before setting the 20 6x6 treated posts that form the core of each column. Each post is screwed to a base bolted into solid blocking below. We braced the posts plumb and then connected them with a box beam (see drawing p. 118). T-straps strengthen the intersection of post and box beam. These posts and box beams set the basic supporting structure of the porch roof.

New Roof with a Pitch

The roof was the final structural part of the porch to build. I suspect that problems with the flat roof over the original porch led to its demise. In the new design we changed the roof to a 3-in-12 shed roof that would drain water easily. On the house side, the rafters were installed on a ledger epoxy-bolted to the brick wall, with the box beam supporting the other ends of the rafters. We extended the eaves to a depth of 3 ft. to allow better rain protection for the tops of the columns and to emphasize the ornamental aspect of the roof edge. Each rafter terminates in an S-curve that tapers to a wedge. This shape echoes the wooden brackets on the gables of the main house.

The roof sheathing went on quickly, and we covered the lower edge with ice-shield membrane and 30-lb. felt paper the rest of the way up. The roof was finished with three-tab asphalt shingles. We sealed the new roof to the building by scoring the brick around the perimeter and inserting copper counterflashing. A beadboard ceiling finished off the roof from below.

Adding the Dressing

With the porch structure basically complete, we turned our attention to the columns, arches, and railings. We didn't want the new porch to suffer the same fate as the old, so each 6x6 post was wrapped with ¾-in. exterior medium-density fiberboard (MDF) that would be the first line of defense against rot (see drawing top p. 119). This MDF skin also turned the posts into 10x10 columns that had the visual weight and heft needed for this size porch.

We built the MDF columns in our shop, assembling three sides in a U-shape that was then slipped around the posts. The final face was biscuited and glued in place to complete the columns. The MDF was sealed on both sides and was held slightly off the porch floor to stop moisture from being trapped.

The tops and bottoms of the columns were finished off with mahogany moldings. Next, we installed the prefabricated arches and keystones between the columns. The arches were made from three layers of mahogany laminated together. A computer-controlled router cut the decorative profiles in each arch. The keystones, which are solely decorative, were made in two halves that lock over and around the top of each arch.

The porch railings were part custom, part stock. We purchased S-shaped Victorian clear-cedar balusters from a nearby railing-parts company. We designed our own railing cap, a wide, gently curved mahogany piece with stepped edges. The railing sheds water but doesn't offer a place for people to put cups.

We used the same railing detail on all three sets of stairs. On each set of stairs we installed an additional pipe handrail that follows the wood railing, making the stairs code-compliant and easier to climb.

Kevin Wilkes is the founder of Princeton Design Guild, a design/build firm in Princeton, New Jersey.

Elegant arches flank an outdoor corridor. The porch provides access to rooms inside the house while creating a spot for a leisurely swing in the hammock on a sunny afternoon.

Adding a Covered Entry

A DDING ARCHITECTURAL DETAILS TO HISTORIC HOUSES HAS been common practice in New England for centuries, but it's always been a tricky business. Building styles change constantly, and early-American design has cycled in and out of fashion. Greek-revival doorways and Italianate porches have been added to 18th-century and early 19th-century homes. In other cases, homeowners have tried to reproduce looks that predate their original houses.

Balancing the owners' tastes with the stylistic requirements of classic New England buildings is often difficult (and sometimes impossible). Luckily on this project, the owners had a great sense of what would work.

My company was asked to add a front-entry portico to a house built in 1826 and located in what is now a historic district of Fairfield, Connecticut. We built the portico on site using mostly stock materials. Our total time on the job, including the design and the construction, was about 90 man-hours.

A new entryway carefully tailored to the original form. The local historic commission stipulated that the curve of the portico frieze match the curve of the fanlight above the front door.

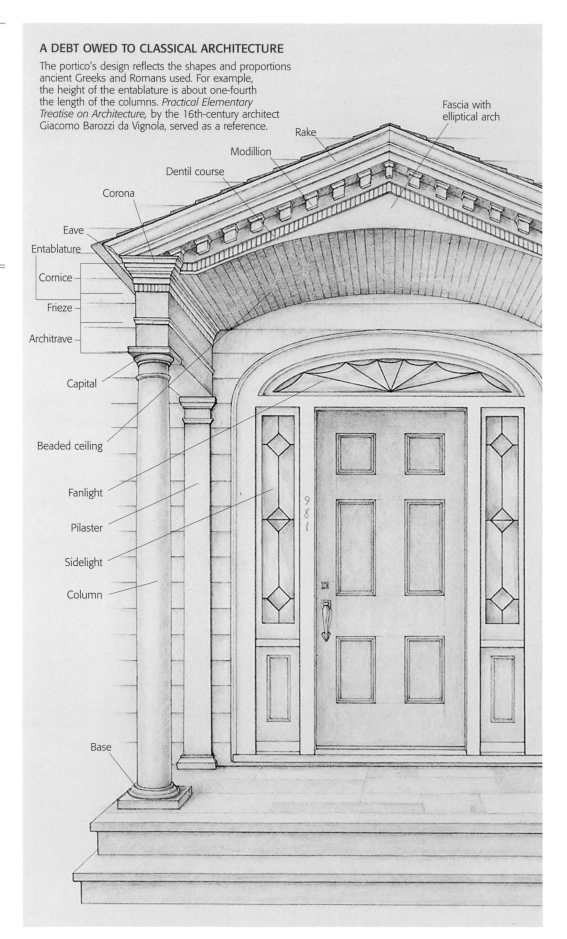

A DEBT OWED TO CLASSICAL ARCHITECTURE

The portico's design reflects the shapes and proportions ancient Greeks and Romans used. For example, the height of the entablature is about one-fourth the length of the columns. *Practical Elementary Treatise on Architecture,* by the 16th-century architect Giacomo Barozzi da Vignola, served as a reference.

Rake

Modillion

Dentil course

Corona

Fascia with elliptical arch

Eave

Entablature

Cornice

Frieze

Architrave

Capital

Beaded ceiling

Fanlight

Pilaster

Sidelight

Column

Base

Starting with a Classic Colonial

The house has the shape and proportions typical of the area's late-18th-century colonials. The five-bay facade faces the street, and two large brick chimneys flank a central stairway and entry hall. Like many early houses in Connecticut, this one had undergone many changes. Paired Italianate brackets supporting a built-out overhang were added to the main roof in the late 1800s, and windows throughout the house were replaced in the early 1900s with six-over-one lite sash.

The front entryway is composed of a six-panel door flanked by leaded sidelights and capped with an elliptical fanlight, all in the Federal-revival style. They were added sometime in the early part of this century.

In deciding how to design and build the new entry portico, we had many factors to consider. There was no evidence in photographs or architectural remnants of what the original portico looked like, if there ever had been one. We had to come up with a design that would fit stylistically and proportionally with the house and at the same time comply with the requirements of the local historic district.

Although there is clear historical evidence that center-hall houses with this type of entry-door surround did not often have projecting porticoes, the owners just as clearly wanted one. We decided to base our design on classical proportions and then modify it to fit our requirements. Builders in the 18th and 19th centuries commonly used this vernacular approach. Their individual expressions of classic designs, I find, often are the best features of early houses.

We presented the final design to the historic district for review. The plan was approved by the local historic commission with the stipulation that the elliptical arch of the fascia on the new portico follow the curve of the fanlight, and that we not attempt to echo the Italianate brackets of the main house in the soffit of the new portico (see drawing facing page). The owners gave the go-ahead to begin building, leaving the exact molding details and proportions to us.

Paul Curtin, my lead carpenter, and I built the portico. Having Paul to bounce ideas off of always makes a complex job go much faster, with fewer mistakes.

Perspective Can Improve the View

We worked out the exact design dimensions on paper, marked the outline on the front of the house with chalklines, and then stepped back to check the proportions of our lines against the facade. This habit of taking the time to stop, step back, and look at what we are doing is one of the most important lessons we have learned from working on old buildings.

After a final check of our measurements and a deep breath, I cut the outline with a circular saw. To our surprise, we discovered an earlier, possibly original wood-shingle layer underneath the top layer of siding, along with the remnants of an earlier doorway. It was an interesting find, but we saw nothing that led us to change our approach. Having cut and removed the shingles, we got a good outline view of the size and proportion of our design.

We decided at this point to rough-frame the entablatures first. This way, we could frame the roof assembly, build the cornice and raking cornice, and get the portico decked and shingled. We would prop it up with 2x4 braces and then install the columns when they arrived.

Taking the time to stop, step back, and look at what we are doing is one of the most important lessons we have learned from working on old buildings.

The author began building the portico by framing the entablatures. The cedar shingles were cut away to accept the new roof flashing (photo left). Curved collar ties made from ¾x12 pine boards support the roof framing (photo right) and serve as backing for the beaded ceiling.

Durable, Weather-Resistant Materials

Our material choices were fairly simple. Given the small roof size, 2x4 Douglas-fir rafters, 16 in. o.c., were adequate, with ½-in. plywood sheathing. Asphalt roof shingles would match the recently reshingled main roof. For durability and rot resistance, we chose redwood for the hollow, turned columns from Hartmann Sanders Co. (see Sources). The column bases and capitals are molded fiberglass, not exactly a traditional material, but durable and indistinguishable from wood when painted.

Historically, these columns would have been made from old-growth eastern white pine. The lumber from these trees was extremely rot-resistant, weathered well, was easy to work, and held paint. Almost all of the antique exterior and interior trim, moldings, clapboard siding, and roof shingles in our area were made from it. We often find 200-year-old exterior-trim pieces

that were maintained poorly, but the old pine still is in pristine condition.

For this job, we milled some of the trim and architectural elements from 1x12 and ¾x12 eastern white pine. We can still find some quality New England pine in our area, good enough for exterior work if it will be kept painted but by no means the equal of the early stuff. For moldings, we used stock pine profiles where possible. All wood was back-primed with oil paint before installation.

The two projecting entablatures were laid out using *Practical Elementary Treatise on Architecture* as a guide. This book, by the 16th-century Italian architect Giacomo Barozzi da Vignola, contains intricate scale drawings and descriptions of the five classical orders of architecture. (This book is no longer in print, but Vignola's work has been reproduced in many forms.) Each order is distinguished by its own proportions and decoration. The proportions of the portico were based roughly on the simplest of the five orders, the Tuscan. Vignola says the height of a Tuscan entablature should be one-quarter of the height of the column supporting it. Using this reference point as our guide, we made the entablature about 20 in. tall. The 2x4 framing for the entablatures was screwed to the house, with the outboard end temporarily propped up (see left photo facing page).

A 2x14 Top Plate

Each entablature was capped with a 2x14 piece of clear pine trimmed with cove molding. This cap piece, the corona, would act as the wall top plate, simplifying the roof construction and permitting a wide, overhanging cornice and drip.

The pitch of the portico roof, 5-in-12, was determined by existing conditions, including the width of the stoop below and the distance to the window above. The rafters were braced with a temporary 2x4 collar tie while we installed the plywood decking.

Taking a cue from traditional building, we carefully ripped the decking at the edges to provide angled nailing support for the back of the crown molding to be installed later. Period roofs usually were sheathed with full 1-in.-thick oak, pine, or chestnut roofers. These boards sometimes ran by the rafters, and their ends were cut on an angle to accept rake-molding nails.

Even Good Carpenters Fudge Things Occasionally

Before we could shingle the portico, we had to complete the cornice moldings at both the rake and the eave. The plan was to use stock 4⅝-in. crown molding. Because the molding on the rake and the molding on the eave are in different planes, these two pieces can be difficult to miter. In this case, the joint was complicated further because

the bottom cove of the eave crown molding was ripped off where it meets the top of the 2x14 corona.

One way to improve the alignment of the rake and eave moldings is to use a 4⅝-in. crown on the rake and a 3½-in. crown on the eave. A more precise way is to have the crown molding for the eaves milled to a slightly different profile so that it matches the rake crown as the two pieces come together at the miter. This is the method the old-timers might have used, and on a bigger job we would have taken this approach. Instead, we adjusted the eave molding by tipping it down and away from the roof slightly and filing it so that the miter appears perfect to the eye.

Once the roof moldings were on and painted, we installed a rubber ice and water barrier over the plywood and over the edge of the crown molding, trimming it with a utility knife. We then stapled down 15-lb. felt paper and, over that, nailed a starter course of red-cedar shingles to serve as the drip edge. A cedar drip edge is slower to install than an aluminum one, but it looks much better on houses such as this one.

Curved Collar Ties Frame the Ceiling

Finally, on went the fiberglass shingles, step flashed with 6-in. by 8-in. pieces of copper that are bent at right angles. We slipped the flashing behind the existing wall shingles after clipping the nails with a reciprocating saw. It was a relief finally to have a roof over our heads, if only in a manner of speaking.

Our plan for the portico called for a high, curved ceiling that left little room for collar ties. With a fairly flat roof pitch and no ties, we were concerned that the roof would spread over time. We solved the problem by designing curved collar ties that also would serve as backing for the curved

Taking a cue from traditional building, we carefully ripped the decking at the edges to provide angled nailing support for the back of the crown molding to be installed later.

Saw kerfs define the dentils.
The dentil molding was milled with
a sliding miter saw set to cut almost
through a pine trim board.

ceiling (see right photo p. 126). One tie for
each pair of rafters was made from ¾x12
white pine. We could not get the entire
curve from the 12-in. boards, so we added
pieces to the bottom of each rib.

With the roof firmly trussed, we were
ready to turn our attention to the various
boards and moldings that make up the
entablature, including the architrave and
frieze. We executed our design, loosely pat-
terned after the profiles and proportions of
the ancient Greeks and Romans, mostly with
stock moldings. Profiles not readily available
were made on site with a router table.

The molding that divides the architrave
and the frieze is not only an architectural
element but also a piece of trim that covers
the joint between two boards.

Site-made Dentil Molding

Following the lead of porticoes on other
houses in the neighborhood, we decided to
add a dentil course along the bottom of the
cornice (see photo above). We made the
dentil molding on site with a sliding saw.
We started by reducing the thickness of a

pine 1x2 to ½ in. Then we placed the stock
flat on the saw table and set the blade so
that it would cut nearly but not quite
through the stock. The ⅛-in. saw kerfs are
about ¾-in. apart.

The dentils were installed below a 1x3
with a ¼-in. round bead at the lower edge.
The resulting dentil course was simple and
fast to install.

Column Profile Used to Lay Out Pilaster

By now the columns had arrived. They were
cut to length and thoroughly primed
according to the manufacturer's specifica-
tions. Column bottoms take a lot of wear
and tear from the weather, so it's important
to seal and install them properly. We
applied a thick bead of polyurethane caulk
to the top and bottom of each column
before attaching the capitals and bases with
screws. We like to use Sika LM polyurethane
caulk at all crucial joints (see Sources). It
stays flexible, moves as wood expands and
contracts, sticks tenaciously to almost anything, and is paintable. Sheet-lead flashing
was placed over the capital before the
columns were put in place, plumbed, and
attached to the roof.

Earlier, before installing the columns, we
used one column as a template to lay out
the two flat pilasters below the portico at
the house. The ancient Greeks knew that
due to an optical illusion, a straight-edged
column appears concave. They used entasis,
a convex curving taper from bottom to top,
to counteract this illusion. Correctly made
columns should exhibit this gentle curve,
which begins one-third of the way up on a
Tuscan column.

We ripped, planed, caulked, and screwed
together the pine pilasters. They were held
plumb against the shingles, and the curving
line was marked out. As before, we cut
through the shingles with a circular saw.

The portico was built to match existing conditions. The dimensions of the portico and the pitch of the portico's roof were determined largely by the width of the stoop and the height of the second-story window above.

Column bottoms take a lot of wear and tear from the weather, so it's important to seal and install them properly.

The pilasters were nailed in place and caulked. We attached the trim here—and throughout the job—using a pneumatic trim gun and 8d galvanized nails.

A Beaded Ceiling Follows the Curve

Period porticoes usually had plastered or paneled wood ceilings. We opted for ½-in. by 4-in. beaded, tongue-and-groove fir wainscot. Available at most lumberyards, it's a quick, attractive solution that looks right on almost any porch. We nailed it up with an 18-ga. brad nailer and 1⅜-in. brads. A quarter-round molding covering the ends of the wainscoting completed the ceiling.

As a last detail, we decided to include modillion brackets running up the raking cornice. We chose a simple shape, known as a block modillion, as opposed to the complex molded brackets that the Greeks or the Romans would have carved. Each block, fabricated to sit plumb on the frieze, is trimmed at the soffit with ½-in. bed molding.

Finally, we patched in the sidewall shingles with red cedar to match the existing, caulked all the joints, and painted the portico with two coats of gloss-white oil paint.

Christopher Wuerth is a restoration contractor in Hamden, Connecticut.

SOURCES

Hartmann Sanders Co.
4340 Bankers Circle
Atlanta, GA 30360
(800) 241-4303

Sika Corp.
201 Polito Ave.
Lyndhurst, NJ 07071
(800) 933-7452

A Shoji Bath

I GREW UP IN ARIZONA, WHERE THE SKY'S LUMINANCE IS VAST.
It was quite a change to move to the Piedmont region of North
Carolina, where a combination of lush trees and a relatively flat landscape limited the visible sky to what could be seen directly overhead. It was
this reduction of daylight that inspired me when my wife and I decided to
remodel our 1950s-style bathroom.

FACING PAGE A small 4-ft. addition yields a tastefully designed master bath with a sky light and modern fixtures.

LEFT Sliding screens offer privacy without loss of light. To avoid wasting floor space, the author used sliding doors made from a water-resistant acrylic. Photo taken at B on floor plan.

SMALL ADDITION YIELDS BIG CHANGES

Faced with a narrow bath entered from a hallway, the author extended the bedroom and bath space by 4 ft. and gained room for a shower under a big skylight. To make the narrow room seem larger, he also closed off the hall entrance and opened a wide doorway from the bedroom side.

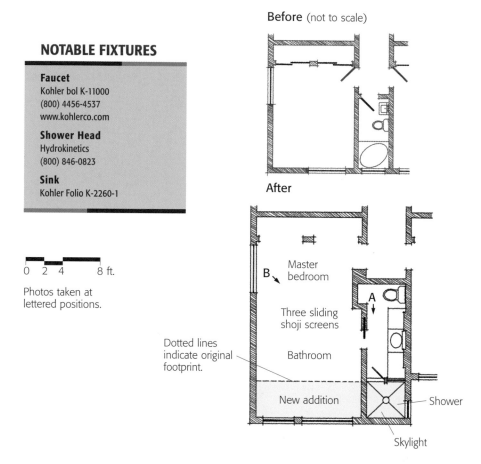

NOTABLE FIXTURES

Faucet
Kohler bol K-11000
(800) 4456-4537
www.kohlerco.com

Shower Head
Hydrokinetics
(800) 846-0823

Sink
Kohler Folio K-2260-1

0 2 4   8 ft.

Photos taken at lettered positions.

Before (not to scale)

After

B → Master bedroom

A

Three sliding shoji screens

Bathroom

Dotted lines indicate original footprint.

New addition

Shower

Skylight

The addition's foundation was a simple combination of footings and concrete-block stemwalls.

A Small Addition Is Enough

Originally, the bath was a 4-ft.-wide space at the end of the narrow hallway that ran to our bedroom. As might be expected, the bathroom was a period piece tiled in robin's-egg blue with turquoise fixtures. A 4-ft.-long oval tub completed the room.

The bath obviously needed a face-lift. Flanked on each side by narrow bedrooms, the bath had nowhere to go but out. I closed off the bath from the hall and extended the bath and bedrooms' exterior wall by 4 ft.; the addition's foundation was a simple combination of footings and concrete-block stemwalls.

By increasing the length of the rooms, I now had more space for a larger vanity, and

I could add a shower illuminated by a large skylight. The skylight would not only capture a glimpse of the changing sky but also would create a well of light contained by the clear, Eurostyle glass enclosure (see photo p. 130).

Sliding Screens Conserve Space

For our convenience, I moved the bath entrance to our bedroom, a move that also made the room feel wide rather than long and narrow. Because the bedroom is also narrow, sliding doors seemed the logical choice to enclose the bath; they wouldn't take up any floor space. Shoji screens would be even better (see photo p. 131) because

Great Idea: Counter Edges That Shed Water

To keep water away from the vulnerable joint between the counter's laminate top and the bullnose edging, cabinetmaker Greg Gates of Charlotte, NC, glued the maple bullnose to the substrate first, then cemented the laminate to the substrate, overlapping the joint by at least ⅜ in. He then routed the edge detail with a ⅜-in. dia. roundover bit, cutting a thin ledge that sheds water away from the glued surfaces. **—D.B.**

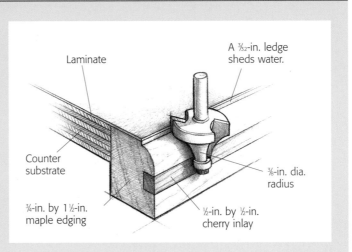

Laminate

A ³⁄₃₂-in. ledge sheds water.

Counter substrate

⅜-in. dia. radius

¾-in. by 1½-in. maple edging

½-in. by ½-in. cherry inlay

---

they could emit a diffuse light into the bedroom when closed, yet still give us some privacy.

Because shoji are traditionally made of rice paper, I was concerned that a misguided hand might puncture the thin panels or that steam from the shower would turn the screens to mush. I avoided the problem with acrylic-modified paper that's tear and moisture resistant from CTT Furniture (see Sources). At $78 per 3-ft. by 6-ft. sheet, the 0.45mm-thick paper wasn't cheap, but it was exactly what I needed. Cabinetmaker Greg Gates built the shojis. To keep the screens in line at the bottom, Greg set them in a grooved oak threshold. The sliding-door hardware is made by Hawa AG (see Sources).

Fixtures Add to the Ambience

To get the most out of the 4-ft. depth of the bath, I designed a vanity 7½ ft. long and 21 in. deep. Taking a space-saving cue from the shoji, the cabinets have sliding doors. Greg used maple plywood overlaid with a cherry grid to repeat the shoji design. He also added a detail on the laminate top that minimizes water damage (see sidebar above).

Dale Brentrup is an architecture professor at the University of North Carolina/Charlotte.

SOURCES

CTT Furniture
San Diego, CA
(858) 587-9311
www.cttfurniture.com

Hawa AG
distributed by Häfele
(336) 889-2322
www.hafeleonline.com

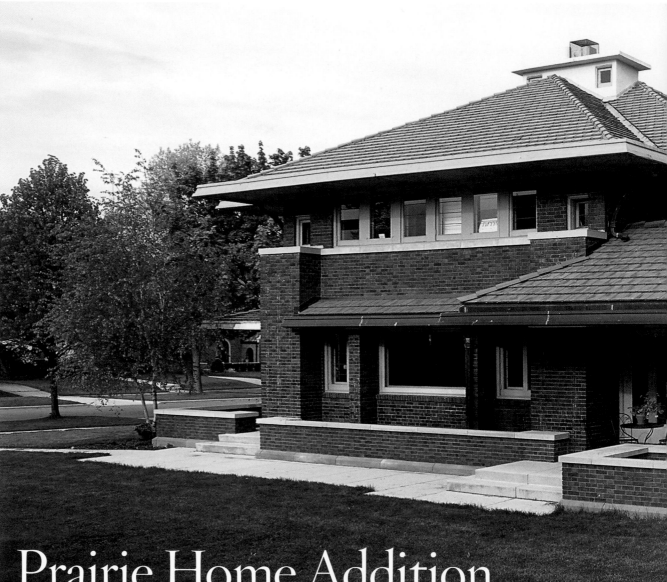

Prairie Home Addition

I N 1988, THOMAS AND MARY PAT CLASEN PURCHASED A 1915
Prairie-style home in Shorewood, Wisconsin, that had become a
showcase of neglect and insensitive alteration. Carpeting covered the
quartersawn white-oak flooring, oak trim had been painted, and red-flocked
wallpaper covered the stairwells. The original Roman-brick fireplace had
been plastered. If only the 1950s hadn't happened.

After the Clasens purchased the home, they set about to restore it to the
original Prairie beauty they knew was hidden beneath those layers of alter-
ations. But the time soon came when the Clasens' needs required more
than the work they could do with their own hands. Wanting to remodel

Prairie-style homes are characterized by strong horizontal lines. The low roof of the addition and the stone-capped brick planters in front accentuate the horizontal and help to merge the new and the old. Photo taken at A on floor plan.

and add to their home, they came to me. The existing home had great formal living spaces but lacked a family room. With four kids, the Clasens needed that informal space.

In addition, the Clasens wanted to remodel the study at the front of the house. There was to be a study desk, lighting, and a computer alcove. The front porch needed replacement. And the access to the side entrance was a wooden deck that wasn't in keeping with the style of the house. That would be changed, and all the remodeling would reflect the original Prairie styling of the house.

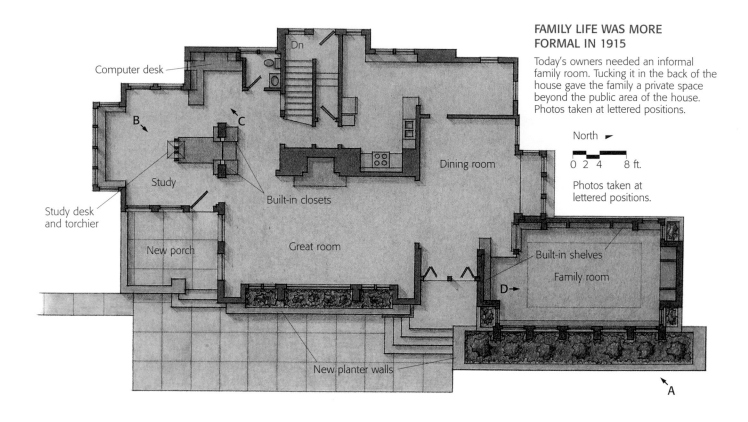

Computer desk

B

Study

Study desk
and torchier

New porch

Built-in closets

Great room

Dn

Dining room

**FAMILY LIFE WAS MORE
FORMAL IN 1915**

Today's owners needed an informal family room. Tucking it in the back of the house gave the family a private space beyond the public area of the house. Photos taken at lettered positions.

North ►

0 2 4   8 ft.

Photos taken at
lettered positions.

Built-in shelves

Family room

D ►

New planter walls

A

The two-story mass of the main house now combines with the long, low lines of the addition and leads the eye out- ward, toward the horizon.

Prairie Houses Are Horizontal

My goal was to bring the Clasen house back to its original beauty, and then add to it in a way that the distinction between old and new was not obvious. I had designed new houses in the Prairie style and studied the style extensively. But successfully blending new to old would be a challenge. The new family-room addition had to connect to the existing house without appearing as a disjointed appendage.

Prairie architecture reflects the long, low expanses of the Midwestern plains, and the shallow hip roof of the Clasen house is a classic example of this tenet. It had to remain the dominant feature, with the family room subordinate to it. We accomplished this by capping the family room with a low, spreading hip roof whose fascia tied into an existing fascia above the living-room windows. Tying the fascias together

links the addition to the house so that the addition looks as if it had always been there. The two-story mass of the main house now combines with the long, low lines of the addition and leads the eye outward, toward the horizon.

I capped the new porch's brick wall with a cut-stone sill and extended this wall to enclose low planters off the living room and the new family room. The planters unify the old and new sections and accentuate the horizontal disposition of the house.

Prairie design uses several visual devices to bring the outside in. Rather than being placed singly on a wall, several windows are combined in bands to give a broad view of the outdoors, a dominant feature of the Clasen house. The new family-room roof cantilevers over the entry to the living room. This sheltered entrance is another characteristic of the Prairie style that softens the distinction between indoors and out.

Selectively Using New Materials

The family-room windows were all custom-sized and produced by Loewen. I thought it better to use a modern, energy-efficient window than to try to copy the existing windows exactly. However, I did have Loewen custom profile the brick molding to match the existing.

Although the original green glazed roof tile is still produced, it was too costly to use on the addition. Also, its unweathered appearance would have made the new tiles stand out from the old. Instead, the builder (Dan Ward), the clients, and I selected a smooth-surfaced, interlocking concrete tile made by Vande Heys Roofing Tile Company (see Sources). Vande Heys custom dyed these tiles to match the existing weathered roof. These new tiles are designed specifically for cold climates. They absorb little water, so they aren't much affected by freeze/thaw cycles.

Stained Glass Lights Up the House

Stained glass is an important element in Prairie houses. It emulates but doesn't imitate natural colors and geometries. Instead, it borrows its forms from nature. For example, noted Prairie School architect Frank Lloyd Wright's sumac design is a composition of triangles that are ordered as sumac leaves are. More than any other interior detail, the stained glass in the Clasen remodel says "Prairie."

The family room's tray ceiling has stained-glass light fixtures set in wooden frames. Stained-glass room dividers and light fixtures add color and sparkle to the otherwise earth-toned house. Oakbrook Esser studios produced the art glass (see Sources). Unlike contemporary art glass, most of the glass in Prairie designs is clear.

The geometric patterns of the caming are integral to Prairie-glass design, and too much color can overpower them. Lighting fixtures are the exception because the lamps behind the glass must be concealed.

The stained glass in the Clasen house uses brass caming rather than the more common lead. This peaked, or colonial, brass caming is then given a dark patina. The result is a much crisper profile than can be achieved with lead.

But Wright Never Designed a Computer Desk

As was typical for Prairie homes, the woodwork in the original house is quartersawn white oak. Quartersawing is more wasteful of wood than is plain sawing. But quarter-

Deep, autumn tones are characteristic of this style, giving these homes a warm, natural feel.

Stained-glass torchier lights the study. The tall, slatted end of the desk forms the torchier's base and fosters needed privacy at a desk only paces from the front door. Photo taken at B on floor plan.

Good renovation blends the new with the old. Details such as stained glass make the computer alcove look to have been built well before its time. Photo taken at C on floor plan.

One remodel highlight is the new desk in the study with its integrated stained-glass torchier lamp (see photo p. 137). Wright sometimes designed furniture to define architectural space in his homes. Some of his early dining-room sets used high-backed chairs to define an intimate space in the larger dining room. Similarly, the woodwork supporting the Clasens' torchier screens the study desk, making it seem more private. Perhaps the most challenging blending of old and new in the Clasen house was the computer desk (see photo left). For obvious reasons, there isn't an original Prairie-style example. But by repeating the molding profiles and observing the shapes of the other cabinets, the computer desk was made to fit right in.

The painters questioned the paint color I suggested for the walls and ceilings because it is much more yellow-gold in hue than is typically used in houses today. I thought the color was necessary for an authentic Prairie feel. Deep, autumn tones are characteristic of this style, giving these homes a warm, natural feel. Wall colors, wood trim, fabrics, and furniture all unite to reinforce this, and the Clasens have found the colors warm and comfortable to live with.

Throughout construction, the Clasens remained true to the vision of unity between old and new. I think we achieved this unity. I have passed the house at night and seen the new stained-glass torchier glowing in the old study. Its shape represents the upward-turned petals of a flower, lending beauty to its surroundings and helping to blend old with new.

Architect Ken Dahlin, founder of Genesis Architecture, LLC, practices in Racine, Wisconsin.

SOURCES

Loewen
600 Lakeview Parkway
Vernon Hills, IL 60061
(847) 362-1600

Oakbrook Esser studios
129 E. Wisconsin Ave.
Oconomowoc, WI 53066
(414) 567-9310

Vande Heys Roofing Tile Company
1565 Bohm Drive
Little Chute, WI 54140
(800) 236-8453

sawing oak reveals a beautiful flecked grain pattern and the boards are more dimensionally stable. We used over 1,400 bd. ft. of quartersawn white oak for this project.

Built-in cabinets (see photo facing page) line three walls of the family room. Their trim mirrors original moldings in the house. Molding cutters were modeled from existing profiles so that new trim would integrate with old woodwork.

Built-ins hide TV and stereo equipment. The stained-glass light on the ceiling and extensive wood trim are Prairie details in a new room that has a modern purpose. Photo taken at D on floor plan.

A Greek Revival Home Gets a New Kitchen

CONTEMPORARY KITCHENS HAVE BECOME TRUE GATHERING rooms where people relax with family and entertain friends. That was not the case among the local gentry of Concord, Massachusetts, when this home was built in 1844 for the Rev. Barzillai Frost, a local minister who was reputedly as cool as his name. In those days, food preparation was harsh work; the original kitchen was a Spartan facility designed for the parson's wife assisted by three maids. The dilemma faced by the present homeowners was how to remodel the kitchen to accommodate a contemporary lifestyle while respecting the antiquity of their home.

Modern amenities such as adequate storage and counter space were always lacking, but a 1950s remodeling had left the kitchen even more dysfunctional than before. At that time, the interior basement stairs were removed. Natural lighting was also severely reduced by the lavatory along the west wall, which blocked the afternoon sun, and by the high windows above the sink counter, which restricted morning light and the view of the backyard garden.

Antique furniture and reconditioned appliances retain the feel of the 19th Century in a kitchen designed to function well into the 21st.

A bad 1950s remodel (was there any other kind?) left the kitchen, sitting room, and lavatory sharing the same dismal, cramped space. Switching the locations of the kitchen and the sitting room, pushing out the south and east walls of the new sitting area, and shoving the old carriage shed backward provided the space and the light to make all these rooms comfortable.

Before

▶
North

Photos taken at lettered positions.

Old carriage shed

Sitting area

Kitchen

Formal dining room

After

Garage (duplicate of old carriage shed)

Back hall

Porch

Dn

Pantry

Historic windows (could not be altered)

D

Kitchen

Hutch

C

E

A

Formal dining room

Sitting area

B

Wet bar

Formal dining room

Euro-Cabinets and Formica Were Out

My clients' primary goal was to create a kitchen that respected the age and style of their home. Toward that end, they researched the elements that would have gone into a 19th-century kitchen, and even interviewed an elderly man who had been born in the house to get his memories of the original layout. Armed with this information, we decided to replicate the feel of a traditional kitchen by installing separate storage units, freestanding appliances and task-specific work areas rather than connecting appliances, and storage areas under a monolithic counter and matching cabinetry.

Along with the kitchen work, my clients wanted to maximize the view of the garden and to create a sitting area for casual entertaining. Their wish list also included a mudroom, basement stairs, a better location for the guest lavatory, a pantry, and a wet bar.

To achieve these goals, we decided to swap locations for the kitchen and the sitting area (see floor plans above). The sitting area would move into an expanded, sun-

Let there be light. Bumping out the south and east walls of the old kitchen creates a spacious and sun-filled sitting area. Photo taken at C on floor plan.

roomlike space on the south wall, facing the garden (see photo above). Bumping out the sitting room would allow the kitchen to fill the northwest corner of the house, freeing enough space for three separate work and storage areas as well as a traditional pantry and a fireplace.

The rest of the wish list was satisfied by creating new space. The old carriage shed would be moved 15 ft. back on the property so that a rear entry hall—complete with closets, basement stairs, and guest lavatory—could be built between it and the kitchen.

Reviews Reinforce Historic Feel

Remodeling in Concord, Massachusetts, is not as simple as it is in Houston, Texas. As our design took shape, we had to have the plans scrutinized by the Concord Historic Districts Commission, which has the final say over any exterior changes visible from a "public way." In our case, the commissioners preferred that the two windows on the north wall of the kitchen (the front wall of the house) not be altered, despite their low sills.

Triple-hung windows with true divided lites fill both exterior walls, providing light and vistas of the garden that can be enjoyed even from the kitchen.

A made-to-order antique. Designed to look like it's been in the family for years, this hutch is the equivalent of an upper and lower cabinet unit (complete with task lighting and electrical outlets). Photo taken at E on floor plan.

Placing counters beneath these windows was out of the question, so we decided to work around them. A new china cabinet was built between the windows to store dishes and glassware and to provide counter space.

Because the house sits on a corner lot, the exterior design of the south wall in the sitting area was subject to review. From the terrace, the sitting area resembles a conservatory. Applied columns between the win-

dows (see photos above and right) were detailed to resemble the two-story-tall pilasters that distinguish the front façade of the house.

The sitting-room windows also create a link to the past. Large custom-made, triple-hung windows with true divided lites fill both exterior walls, providing light and vistas of the garden that can be enjoyed even from the kitchen. Unfortunately, the carriage shed was unable to be saved, but its shell was reproduced exactly by Hampton Smith, the historically attuned contractor, and now functions as an attached one-car garage.

A new space that fits an old house. Wide frieze bands and applied columns (see photo above top, taken at B on floor plan) pay homage to the elements that identify the Barzillai Frost house (see photo above bottom) as an American Greek revival.

A Greek Revival Home Gets a New Kitchen **145**

Well-lighted and convenient storage. Besides shielding a modern refrigerator from the eyes of purists, the pantry has plenty of shelf space to help preserve the uncluttered look of the kitchen. Photo taken at D on floor plan.

Nooks and Crannies Add Charm

When this house was built, the icebox would have been tucked away inside a pantry, not in the middle of a hot kitchen where the ice would not last. So we hid a refrigerator in the new pantry, but placed it just inside the door (see photo above).

Reconstructing the basement stairs in the addition just off the kitchen provided handy storage along one sidewall, making it easy to store infrequently used items outside the kitchen. On the east wall of the kitchen, an old chimney flue that might have served a coal stove was rehabilitated and refitted with a gas-fireplace insert. Tucked into an alcove within a new passageway created between the kitchen and the formal dining room is the wet bar, complete with copper sink, mini-refrigerator, and glass cabinets.

Holly B. Cratsley is the principal of Nashawtuc Architects Inc., an architecture firm in Concord, Massachusetts. She has been designing new homes, renovations, and additions for 22 years. She also provides architectural services to Habitat for Humanity and local shelters.

Functional Antiques Add to Historical Feel

The owners were fortunate to be able to assemble many antiques, from family and other sources, that make this space feel like a much older kitchen. One of these items is a fully operational flour sifter, mounted on the pantry's back wall (see photo top right). Another antique is the kitchen table, built by the wife's great-grandfather. An electrified 1870s oil lantern serves as the central light over the table (see photo middle right). The lantern's support mechanism allows it to be raised out of the way for kneading bread or lowered for meals; other light fixtures were found in the attic and retrofitted.

The centerpiece of the kitchen is a 1928 Magic Chef oven (see photo bottom right), which was restored and brought up to date by David Erickson of Littleton, Massachusetts, a founder of the national Antique Stove Association. More than just a work of art, the Magic Chef provides amenities such as a warming oven, two full cooking ovens, and a waist-high broiler, in addition to a spice shelf and a built-in light fixture. A stained-glass window found in an antique store is built into the west wall above the cookstove. The afternoon sun lets this window cast color through the kitchen. All these antiques were assembled well in advance so that they could be planned into the kitchen.– **H.B.C.**

A Victorian
Blossoms in the Orchard

NESTLED IN AN APPLE ORCHARD, THE HOUSE LOOKED worn beyond its 20 years. The roof leaked, the windows were in bad shape, and the lawn grew long around a swimming pool half-filled with algae-thickened frog water. But as the real-estate folks like to say, it had lots of potential. The house's modest size and Victorian lines appealed to the new owner, Michelle, who saw the house as a great start. Luckily for Michelle, her new next-door neighbor was Roland Batten, the architect who had designed the house; she hired him to design its newest incarnation. The trick was to increase the size of the house without destroying its character.

By the time Michelle bought it, the house consisted of a two-story rectangle with a small addition. The new scheme called for dormers in each of the three upstairs bedrooms, which would add needed space and natural light for the kids without altering the overall feeling of a Victorian cottage (see photo above).

Dormer goes in front, music room goes in back. Architect Roland Batten remodeled his own design, adding a small addition and dormers to the front of the house (photo above taken at A on floor plan). A music room in the rear (photo facing page taken at B on floor plan) features an arched roof and wavy shingles.

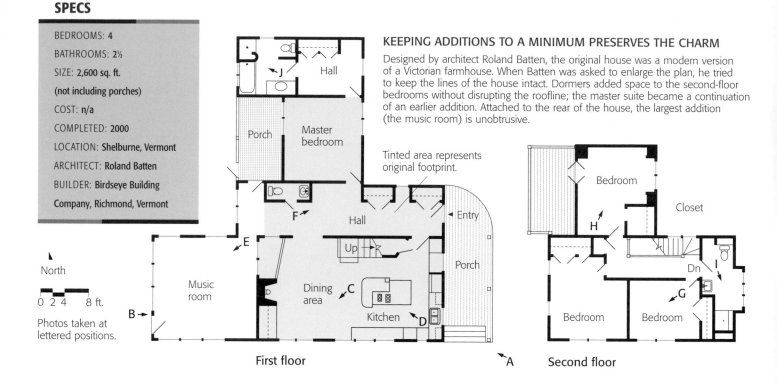

North

0 2 4   8 ft.

Photos taken at
lettered positions.

KEEPING ADDITIONS TO A MINIMUM PRESERVES THE CHARM

Designed by architect Roland Batten, the original house was a modern version of a Victorian farmhouse. When Batten was asked to enlarge the plan, he tried to keep the lines of the house intact. Dormers added space to the second-floor bedrooms without disrupting the roofline; the master suite became a continuation of an earlier addition. Attached to the rear of the house, the largest addition (the music room) is unobtrusive.

Tinted area represents original footprint.

First floor

Second floor

The new plan also called for two new rooms. A ground-floor master suite now extends from the north side of the house (see floor plans above). It carries the same Victorian style as the original. The music room, on the other hand, is a journey into a more recent century. Overlooking the formerly murky swimming pool on the south side of the house (see photo p. 149), the music room is topped with a vaulted roof of standing-seam copper and is sided with wavy courses of shingles.

Fix the Leaks First

The house needed repair in some places and rebuilding in others. The original corrugated-metal roofing was leaky, and the finger-jointed windowsills and exterior doors and jambs were all in bad shape. All interior walls were to be reconfigured, and existing electrical, plumbing, and heating needed upgrading.

The dirty work began as we dug a trench along the foundation to find out what was wrong with the footing drains. The original builder had simply put back the soil that came out of the trench. Unfortunately, the local soil is heavy, dense clay, and over time it made the perforated footing drains collapse. This eventually caused the basement to flood. We replaced the drains, and we backfilled with 1½-in. washed stone. So far, that solved the problem.

After discarding the mouse-infested fiberglass insulation, drywall debris, painted lumber, and delaminated plywood, we reused as much 2x lumber as possible. Untreated lumber that could not be reused was loaded in a separate container and taken to a nearby wood-fired electricity-generating plant.

Variety of countertops makes a versatile working kitchen. Occupying one end of the main living space, the two-tiered kitchen island features maple and stainless-steel counters, each for a specific task. Photo taken at D on floor plan (prior to installation of a proper handrail).

Kitchen Mixes Custom and Factory-Built Parts

The new kitchen is a mix of factory-built birch cabinets and custom countertops (see photo above). The kitchen island has two separate levels. The lower counter of maple butcher block surrounds the stovetop burners. A custom stainless-steel top made at a local metal shop steps up from the maple top to bar-stool height. We used a brushed finish that was better at hiding scratches than a polished stainless surface.

We ordered factory-built cabinets to save money for the counters. But it just didn't work out that way. Any apparent savings withered away as the carpenters tried to install improperly ordered and/or carelessly built boxes.

The wood beams and turned vertical-grain fir columns provide a warm complement to the steel counters. In the space that would normally be used for additional kitchen cabinets, a stacked washer/dryer

ABOVE Dining or reading? Why not both? Opposite the kitchen, the dining area is the place to relax. Built-in seating and bookshelves make the space popular throughout the day. Photo Itaken at C on floor plan.

unit provides efficient access for laundry chores.

At the other end of the room, a dining area is strategically positioned between the fireplace and a pair of windows that look toward the orchard (see photo above). On the other side of the fireplace, a pair of interior windows overlook the music room (see photo facing page). The windows provide another view of the orchard from the kitchen, and they also help to tie these two public rooms into a friendly commingling of spaces. A daybed completes this corner of the kitchen.

Winterproofing the House

Keeping a house warm is a three-season chore in Vermont, so we made sure the heating system was up to the task. We replaced the tiny original boiler with a Weil-McLain boiler (see Sources) that could handle both heat for the newly enlarged house and a heat exchanger for the pool. The first floor is heated by radiant-heat tubing in aluminum reflector pans stapled to the underside of the subfloor, and the second floor is heated by Buderus panel radiators (see Sources). Michelle liked the look of the panels, and they were considerably cheaper

Big windows put the music room in the orchard. Taking advantage of the southwestern exposure, the music room's expanse of windows and barrel-vaulted ceiling create a light, airy space in which to practice or perform. Photo taken at E on floor plan.

to install than radiant tubing. The panels come in various sizes to accommodate different room sizes and heat loads.

A good heating system is effective only with good insulation. We've had good results with sprayed Icynene foam (see Sources), especially in renovations. In our area, it's more expensive than fiberglass or cellulose, but it seems to have some real benefits. Applied as a liquid, the foam

expands and theoretically fills every crack, stopping air infiltration. We also have found that it can push through horizontal cracks in walls and roof sheathing to form bulges beneath housewrap or felt paper, a problem easily cured with utility-knife surgery.

Defining a Room's Personality with Color

Painter Polly Wood-Hollend decorated the spot in the hallway below where the owner likes to play her harp. The French-inspired painting features a poem by Yeats. Photo below taken at F on floor plan.

The children chose the paint colors of their rooms. In the orange bedroom, a three-panel mirror seems to make the space larger and echoes the angles created by the dormers.

The uniform green of the upstairs bath makes a good background for the clear-finished vanity and heart-pine countertop.

Teddy's color of choice was purple. Photo top right taken at G on floor plan. **Sonja's orange room.** Photo middle right taken at H on floor plan. **A green bath warms the upstairs.** Photo bottom taken at I on floor plan.

Throwing curves in the bath.
The owner and builders collaborated to design the curved wall that separates the tub from the shower and serves as a shower curtain. Photo taken at J on floor plan.

The Advantages of a Colorful House

The house's interior is brightly painted: warm, colorful, playful, bold, overpowering, or subtle, depending on the room.

Choosing a rosy pink downstairs, a sky blue in the music room, and a warm green in the bathroom, Michelle tackled the selection of paint colors with the same enthusiasm, creativity, and trust that she applied to all the seemingly endless decisions that a homeowner must make. Michelle's kids even got into the act and were given free rein to choose their own colors for their bedrooms (see sidebar facing page). In addition to the regular house painters, Michelle also hired Polly Wood-Hollend to paint a section of

the hall (see left photo facing page) making a colorful alcove where she plays her harp.

Michelle's creativity also spurred the design of the master bath's tiled shower (see photo above). Capturing one end of a large soaking tub, a tiled wall extends past the outer edge of the tub and then curls away, creating the shower enclosure. We framed the wall with the straightest 2x4 studs we could find and put them a mere 6 in. apart to create a tight curve. Multiple coats of plaster were troweled over metal lath until the surface was smooth, fair, and ready for tile.

John Seibert is a partner in Birdseye Building Company of Richmond, Vermont. His firm designs and builds custom homes, cabinetry, and furniture.

SOURCES

Weil-McLain (GV-5 LP gas boiler)
(219) 879-6561
www.weil-mclain.com

Buderus (panel radiators)
(800) 283-3787
www.buderus.net

Icynene (spray-in foam insulation)
(888) 946-7325
www.icynene.com

Adding a Sunroom with Porch

Lots of light for all seasons. When cold weather settles into a long, dark winter, this sunroom is cozy and full of light.

I 'VE SEEN TOO MANY OLD HOMES RUINED BY INAPPROPRIATE additions. In some of those cases, it would have been better to tear down the house rather than graft an unsuitable addition to it. I'm not saying an addition has to be a slavish copy of the original. But it must fit. It must add to the original, not just be added to it.

So when I was asked to design a sunroom and porch addition for a classic Victorian in Montreal, the challenge was to make the addition appropriate to the house and useful to the clients. By the time I met Dennis and Suzanne, I was already familiar with their house, which is one of the few examples of late Victorian in the neighborhood. On the back of their house, which faces west and the couple's lovely garden, a small two-story shed addition had reached the end of its life span (see photo above). The shed was a light frame structure with single-pane windows and, of course, no insulation. The enclosed lower portion of the addition was used for storage. The upstairs balcony was accessible from the second floor.

Dennis and Suzanne wanted to replace the shed with a Victorian-style sunroom covered with a second-story porch. They wanted a powder room in the addition, a full-height basement beneath it, and outside access to the basement, which already was accessible from the main basement. They also wanted the addition to be as maintenance-free as possible.

A fitting replacement for a tired addition. An outdoor porch for warm weather opens to the upstairs hall of the main house, while below it the sunroom is a comfortable setting for all weather.

The Design Had to Suit the House

I decided the sunroom should be designed like a porch, supported by classical columns yet enclosed with large windows and molded panels on three sides. In the center, French doors would open onto a stair leading to the garden.

To prevent obstructions to the view from the addition, we would put the powder room in the corner of the sunroom against the house. To respect the exterior design and the continuity of windows, columns, and molded panels, the powder-room partition would intersect the exterior wall between two windows so that it remained invisible from outside.

To help relate the addition to the house, the overhang above the sunroom would be supported by wide brackets, similar to the brackets beneath the cornice returns on the main house. Victorian brackets and moldings also would enhance the design for windows, transoms, and second-floor columns.

Finally, transoms above the windows and French doors—each topped by a molded panel—would create the illusion of height necessary to match the existing house and to keep all openings in proportion.

The most significant change over the old shed was the enlargement of the basic footprint. The previous 20-ft. by 12-ft. sunroom with four windows and a patio door on the facade became a 27-ft. by 14-ft. structure with six windows across the facade—three on each side of the French doors—and four windows on each side (see photo facing page). Also, instead of a stair running perpendicular to the sunroom as in the old addition, we designed a small landing in front of the French doors with steps coming off each side leading to the garden. The result was only slightly more expensive but much more elegant.

WATERPROOFING THE JOINT BETWEEN NEW AND OLD FOUNDATIONS

A PVC spline spans the joint between new foundation and old. To install the spline, two vertical 4-in.-deep kerfs were cut in the old foundation wall and the thin section between the kerfs removed. The PVC spline was grouted into the enlarged kerf using epoxy concrete. To help ensure a watertight joint, 1x1 keys were nailed to the inside of the forms where they met the old wall. When the forms and the keys were removed, the keyways were filled with epoxy concrete.

Existing wall

Vertical ribbed PVC spline set into old wall with epoxy concrete

2-ft. long, 3/4-in. steel rods at 12 in. o. c.

1x1 keyway, filled with epoxy concrete when form removed

New wall

Ensuring a Watertight Foundation

The new basement wall would be 10 in. thick and have an average height of 5 ft. above the finished grade. Because we needed 7 ft. of headroom in the basement, the concrete slab was placed 2 ft. below grade. To protect the footings from frost, they sit 2 ft. 6 in. below the level of the slab, or 4 ft. 6 in. below grade, which is the minimum depth in Montreal.

SOURCES

Sternson (Durajoint waterstop)
Box 130
Bantford, ON, Canada N3T 5N1
(519) 759-6600
info@stenson.com

Cultured Stone Products
Napa, CA 94559-0270
(800) 255-1727
www.culturedstone.com

Fypon Molded Millwork
Stewartstown, PA 17363
(717) 993-2593
www.fypon.com

M. Q. Doors and Windows
Dania, FL 33004
(954) 929-8500
mqwindows@mindspring.com

To provide a tight, dry connection between the new and old foundation walls, we cut a 4-in.-deep vertical groove in the old stone wall, into which we installed a PVC spline (see drawing p. 159). The ribbed spline, called a Durajoint waterstop (see Sources), was grouted in place with epoxy concrete, which was allowed to harden before we poured the new foundation over it. With the 9-in.-wide waterstop embedded halfway into both the old and new foundation walls, I thought the joint would be watertight.

We also installed hooked steel rods between the two walls—2-ft. lengths of ¾-in. rebar at 12 in. o.c.—which ran beside the flexible spline the height of the wall. We nailed 1x1 strips to the inside of both interior and exterior wall forms. After removing the forms, we filled the 1x1 spaces with epoxy concrete. This measure would provide even greater insurance against water intrusion.

Protecting the Sunroom from Heat and Moisture

The summer porch on the second floor was open to the weather, so we had to waterproof its deck carefully to protect the sunroom below (see photo p. 156). The deck system consists of ⅝-in. pressure-treated plywood sheathing nailed to the sunroom's ceiling joists, then covered with the built-up asphalt roofing.

Perpendicular to the house, wide strips of foam mudsill gasket material run over the roofing every 16 in. o.c.; over that, we set 2x4 sleepers, which are held in place by the weight of the decking. The foam would protect the membrane. We cut slots across the underside of each sleeper at 16 in. o.c. to help the water move in every direction. Finally, the 2x6 treated decking was nailed across the sleepers and spaced about ⅜ in. apart.

To secure a watertight joint between the deck and the house, we raked out the mortar joints between the bricks, then glued and nailed a piece of ½-in. plywood to the house. After the asphalt membrane was installed up and over the plywood, we embedded horizontal aluminum flashing in the mortar joints to protect the edge of the membrane, which rides 6 in. up the masonry wall.

A Hip Roof with Skylights

The choice of a hip roof was evident from the beginning of the preliminary design. An attic window in the gable of the house overlooking the backyard prevented us from building the addition too high. Besides, given the size of the addition, a gable roof would have appeared too tall for the house.

Before construction on the addition began, the clients decided to reroof their house with blue metal standing-seam roofing, which we used to roof the addition. The new roofing makes a strong visual tie between the new and the old.

Originally, I planned to have a flat ceiling over the sun porch made of ¾-in. tongue-and-groove pine beadboard. However, we decided to take advantage of the hip-roof slopes and to fasten the porch ceiling directly to the rafters, adding four skylights, two in the center hip and one on each side. That change provides a much more interesting look inside the porch. Each skylight measures 2 ft. by 6 ft. and provides plenty of additional light on the deck (see photo facing page).

Man-Made Details Save Time and Money

The exterior finish was important to the project. For the large exposed foundation walls, natural-stone veneer was too expensive, and neither stucco nor brick was satisfying. The solution was cultured stone by

Solitude on the porch. Four skylights open the porch to sunlight; a deck over a waterproof membrane protects the sunroom below.

Cultured Stone Products (see Sources). This artificial stone is much lighter and cheaper than natural stone. Natural stone would have been about $20 per sq. ft. The cultured stone cost about $12 per sq. ft.

All molded panels and brackets were made by Fypon (see Sources). This company offers a wide selection of molded millwork made from high-density polymer that can be used like wood millwork, but it comes at a much lower price.

Dennis and Suzanne wanted the addition to be as maintenance-free as possible, so they asked for a balustrade that was made of PVC. I wasn't crazy about the idea, but the difference between wood and plastic is almost invisible. I usually prefer painted wood to PVC, but this balustrade really looks like a wooden one. For architectural unity, the handrail and the balusters along the stairs leading to the garden were made of the same material.

When Dennis and Suzanne replaced existing windows of the house, they chose vinyl windows from M. Q. Doors and Windows (see Sources). Made in Canada with German technology, the windows—unlike most vinyl products—have the beautiful proportions of wood windows, but they are almost maintenance-free. We decided to use the same type of doors and windows for the sunroom. We chose casement windows with double-insulated glass and real dividers. The French doors also have real divided lites.

Didier Ayel is an architect in Montreal, Canada.

For architectural unity, the handrail and the balusters along the stairs leading to the garden were made of the same material.

Kitchen Remodel, Family Style

We had a two-year-old daughter and a baby on the way when we tore up our house and remodeled the kitchen. Not the ideal time to start a project such as this one, but George McNeilly, a contractor with whom I had worked on several satisfying projects, finally had an opening in his schedule. As an architect, I know how important it is to work with a good builder, and I knew George as an outstanding contractor. Besides, my growing family and I really needed more from our old kitchen. Darkened and cramped by its windowless south wall, which also shut us off from our small but appealing backyard, this kitchen had to be opened up.

Among other goals, we needed to open the wall and let in the view of the garden. We also wanted to preserve the character of our house, which has a turn-of-the-century Arts and Crafts feel to it. Our budget was small by today's standards—about $36,000. We agreed to spend our hard-earned savings on achieving these priorities: natural light, a kitchen work area, accessible storage, a play/homework space for our children, and convenient access to our backyard.

Escape to the Backyard

Why do New England houses trap occupants inside? That question was on our minds when we looked at the back wall of our existing kitchen (see photo top p. 164). This configuration gave way to our small addition. We kept this new space to 6 ft. by 13 ft., which saved our backyard apple tree and enough money to finish the inside.

Removing the wall and adding a small, sunny room remedied the problem. Photo taken at A on floor plan.

RIGHT **The job starts here.** The south wall of the old kitchen stood between the daylight and the backyard.

RIGHT **The addition is mostly windows.** A French door opens onto the landing of the addition, which includes a built-in bench/privacy railing and a stair to the backyard that doubles as seating. Photo taken at B on floor plan.

> *Because we can now see it, the backyard has become a significant part of our living space.*

The addition is mostly windows, so it brings in natural light. It also provides space for a kitchen table. Because we can now see it, the backyard has become a significant part of our living space. And we can reach the yard through a door to a new exterior landing (see bottom photo above).

Finding New Uses for Old Parts

The existing kitchen was one dingy room with a nasty metal sink cabinet. But the cabinet doors in the pantry were still in fine shape, and some of them included beautiful panels of ripply old glass. We decided to remove the existing doors carefully, give them a coat or two of paint, and recycle them into the new kitchen. This move would save us a little money and keep some viable materials out of the landfill.

I drew up plans, elevations, and details to incorporate each of our old cabinet doors in new cabinets. In the finished kitchen,

Interior windows add a sense of expanded space. The original vestibule, now also a pantry, lies beyond the breakfast table behind a pair of recycled windows. Photo taken at C on floor plan.

Encore for the kitchen sink. The original kitchen-sink cabinet is now topped with a stock sink in a custom stainless-steel counter. Photo taken at D on floor plan.

I concentrated on storage areas, believing that having a place for almost everything would organize the chaos right out of our lives.

they are surrounded by open shelving, which costs considerably less than cabinets. The open shelves, along with a continuous shelf that encircles the room at cabinet-top height, helps to tie disparate storage elements together visually. What we saved on cabinets we blew on detail and craftsmanship, as I designed various shelf-support brackets and a wood enclosure for new baseboard heat (see photo p. 165).

Peeling away the old linoleum revealed a decent maple strip floor, which we had refinished. That led us to finish the new addition with maple flooring, tinted with a light stain to blend the two better where they meet. New stock tongue-and-groove wainscoting, similar in style and dimension

to the original, replicates the finishes of the existing house in the new construction. Similar is close enough in such a case, and it fits the budget.

When it came to the sink cabinet, we parted with the stained and chipped cast-iron sink top. But the base looked like it still had some life in it, including a lot of useful drawers. We decided to have it professionally painted and had a new stainless-steel top made for it. My mother even got into the act by making new door and drawer pulls out of stock wooden molding that she painted black (see photo above). The work transformed this eyesore into a functional vintage piece.

GAINING A LITTLE SPACE WHERE IT'S MOST NEEDED

An addition of about 80 sq. ft. off the back of this kitchen made room for the work and storage areas to be reorganized with a minimum of structural changes to the interior. The former vestibule remains as the new pantry, while a new door and windows to the backyard connect the kitchen to the outdoors.

Photos taken at lettered positions.

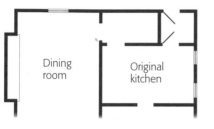

Before

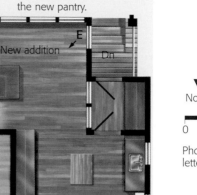

Old vestibule is the new pantry.

New addition

Dining room

After

North

0 2 4 8 ft.

Photos taken at lettered positions.

What, No Dishwasher?

We opted not to earn that Ph.D. in "Appliances: State of the Art." Instead, we chose the basic midline refrigerator and range we were examining at the precise moment that our daughter dropped to the showroom floor in a fit of anguished boredom. We got lucky with the Amana range (see Sources). We really like its heavy burner grates, which don't slide around when we move pots on or off the stove.

We continue to amaze friends and family alike over our deliberate rejection of installing a dishwasher to spend our money elsewhere. We had never owned one and reasoned we would not suffer. I guess renovations are a series of personal decisions and, at times, require restraint.

I concentrated on storage areas, believing that having a place for almost everything would organize the chaos right out of our lives. I can now say that specific storage areas are a big help. For example, in the vestibule we added pantry cabinets and low hooks for our daughters' coats. We reworked the walls of this vestibule, using interior glass doors and windows recycled from a friend's renovation project (see photo p. 165). This decision helped to justify the space taken by the vestibule, as its transparency provides light and a sense of expanded space.

Around the corner from the cooking area is a place for the real work to be done. As I prepare meals at the island by the sink, I spell out loud while our daughters learn to write at an office tailored specifically to their size. Art supplies are stored in a child-accessible lower cabinet.

This project was not inexpensive for us, but 3½ years later, we are more than satisfied. We voted for high-quality craftsmanship, and it brings daily satisfaction. We agreed early on that we were not interested in expensive, precious materials and finishes, but in applying practical, common ones in ways that complement the character of our house.

Architect Ann Finnerty is based in Boston, Massachusetts.

SOURCES

LIGHTING
Illuminating Experiences
Early Electric Collection #40
(800) 734-5858

SINK
Elkay Lustertone
(630) 574-8484
www.elkay.com

RANGE
Amana gas range (ARG7300)
(800) 843-0304
www.amana.com

The porch is part of a composition. The porch includes a stone patio and an octagonal, covered entry. These elements don't just look good, they help keep water away from the porch. In addition, the porch's roof features large overhangs and gutters for protection from sun and rain. Posts tapered to suggest the shapes of tree trunks support the gable overhang. Lights in the soffits shine on the roof and down around the porch perimeter. **Scaffold provides work station for assembling trusses on site.** Heavy 6x8 beams spanning from the house to the porch's corner posts (see photo right) carry trusses constructed of 2x4s and custom-made steel plates. Heavy beams allow for open walls that are sheltered by 2-ft. deep eave overhangs.

A Screen Porch
Dresses up a Ranch

"SOMETHING THERE IS THAT DOESN'T LOVE A WALL; that wants it down." I know this line from Robert Frost's "Mending Wall" evokes a deeper meaning; however, it seems that this sentiment applies whenever man endeavors to build a sound and lasting structure. There is always some element of nature or turn of fashion that conspires to alter the structure once it is completed.

I wonder what Frost would have written had he been building a wooden structure. I know I had my work cut out for me when I added a screen porch to a ranch house in Connecticut (see photo facing page). My clients requested a shaded outdoor-living area that provided views and took advantage of cool breezes coming from the wooded area behind their house. The catch was that the outdoor-living area should have a finished wood floor and lots of wood trim. Digging the foundation 42 in. below grade to get under the frost line was only part of the solution. I designed and built a porch to survive the elements as it provides a beautiful refuge from summer heat and pesky bugs.

Cypress Frame over Crushed Stone

When planning the porch, I followed basic moisture-control strategies: allowing for drainage, creating ventilation, and keeping wood out of direct contact with the ground. During the beginning stages of the project, work was required on the house's septic system. With the backhoe on site, I took the opportunity to cut the grade down about 2 ft. in the porch area. I also sloped the grade away from the house.

With the backhoe I then dug six 42-in.-deep holes for the concrete foundation piers. Once the piers were in place, I bolted triple 2x10 beams around the perimeter of the porch, supported by the piers and bolted securely to them.

Taking the grade down 2 ft. allowed me to add a layer of crushed stone about 3 in. deep and still have almost a foot of clearance beneath the porch's floor joists. The crushed stone creates a clean and well-drained area under the floor.

The floor joists are 2x8 cypress, which costs a bit more than pressure-treated pine but is naturally, as opposed to chemically, rot resistant. Spaced on 2-ft. centers, the joists are connected with joist hangers to the perimeter beams and to a triple 2x10 center-span beam.

To prevent small animals from getting under the floor system, I attached galvanized-steel wire mesh to the inside surface of the perimeter beams, bent the wire back about 6 in. along the ground and covered the wire with crushed stone. During this phase of construction I also prepared the site for a stone patio. The patio encircles the porch and extends slightly beneath it; the floor framing is about 3 in. above the perimeter stonework. Unlike the soil and plants that often surround porches, the stone patio allows rainwater to drain away from the porch, giving it a better shot at staying dry and ventilated. The patio also looks great and helps unify the new construction and the existing house.

Another basic moisture-control measure I took was to provide air circulation underneath the porch. I faced the perimeter beams with a cedar skirtboard that's ¾ in. above the stone. This ¾-in. gap runs continuously around the porch and lets air flow freely through the floor system.

Continuous Headers and Braced Posts Make for Open Walls

After putting single 2x4 pilasters at the house and double 2x4 posts at the corners of the floor frame, I installed two 24-ft.-long 6x8 header beams on top of the pilasters and posts. The header beams run continuously from the house out to their free ends, which cantilever 2 ft. 6 in. beyond the freestanding posts that rest on the stone patio.

I wanted to keep the sidewalls of the porch open to catch summer breezes and to capture the view of the woods as clearly as I could. So I used single 2x4 posts on 5-ft. centers, which would later be fully cased in 1x trim, to frame the screen openings. Single 2x4 blocks at chair-rail height stabilize the posts against bending and twisting and make the finished screen sizes large but manageable.

The continuous 6x8 headers solidly connect the individual posts together at their tops and create a stable and rigid frame out of a row of rather slender posts. The casing of the posts in 1x trim and the horizontal bracing provided by the chair-rail blocks allow the single 2x4 posts to bear the necessary roof loads easily. To add to the rigidity of these sidewall frames, I used sections of 3-in. steel angle with 8d nails to connect the corner posts to the floor system and to the header beams. These angle brackets

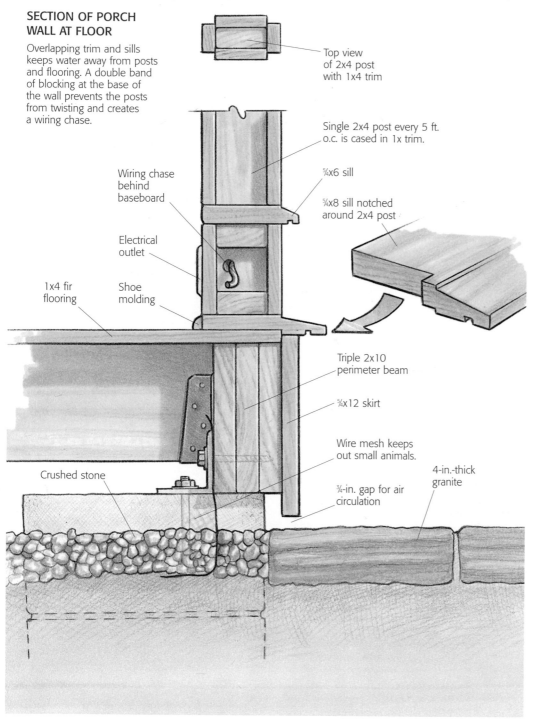

SECTION OF PORCH WALL AT FLOOR

Overlapping trim and sills keeps water away from posts and flooring. A double band of blocking at the base of the wall prevents the posts from twisting and creates a wiring chase.

Top view of 2x4 post with 1x4 trim

Single 2x4 post every 5 ft. o.c. is cased in 1x trim.

Wiring chase behind baseboard

¾x6 sill

¾x8 sill notched around 2x4 post

Electrical outlet

1x4 fir flooring

Shoe molding

Triple 2x10 perimeter beam

¾x12 skirt

Wire mesh keeps out small animals.

Crushed stone

¾-in. gap for air circulation

4-in.-thick granite

To prevent small animals from getting under the floor system, I attached galvanized-steel wire mesh to the inside surface of the perimeter beams.

were later hidden underneath the finish trim work.

I could have used a bottom plate, but instead I opted to install the 2x4 posts directly on the perimeter beams (see drawing above). I laid the flooring on the beams and capped the end grain with a ¾x8 sill that's notched around the 2x4 posts. A double band of blocking topped with a ¾x6 sill creates a wiring chase that, when covered with 1x4 trim, provides an attractive baseboard detail.

Large screen openings celebrate fall colors. Tinted marine varnish on the fir flooring and decay-resistant cedar and cypress trim, primed on all sides, allow for interior-quality details in the screen porch. The trusses, made of small-dimension lumber, create a branch-like effect. In the soffit bays, 1x6 trim boards hide spotlights.

The Site-Built Trusses Create a Branchlike Effect

Inside the porch, I wanted the roof structure to appear light and open, almost treelike in its framework. My idea was to link the structure to the lacework of woods that the porch overlooks. So I designed a cathedral ceiling that would be supported by trusses built with small-dimension lumber, which evoked the image of tree branches better than a framework of rafters and collar ties.

I fabricated the trusses on site using 2x4 fir studs and ⅛-in. galvanized-steel plates cut with a jigsaw. After making patterns, I cut all the truss pieces on the ground and assembled the trusses in place, working from a scaffold. The trusses are on 5-ft. centers.

Between the trusses I installed doubled 2x4 rafters. The rafters are braced at the ridge and at midspan by 2x4 kickers nailed to the trusses. Fanning out from the trusses, the kickers provided the branchlike effect that I had been pursuing.

Installed with the bevel face down, the 2x6 tongue-and-groove roof boards provide a little texture, and the dark lines of the bevel joints look good with the 2x4 framework.

Overhangs Keep Porch Cool and Dry

The porch roof overhangs the walls all around for shade and for rain protection. At the eaves, 2-ft.-deep overhangs match the overhang depth on the existing house. The truss design allowed me to build these overhangs without using large collar ties. Such large members would have appeared much too heavy for the look I wanted. The trusses transfer all of the roof weight to the 6x8 headers, leaving the truss ends free to create the overhangs.

The large soffit overhangs also turned out to be a good place to locate floodlights to light the interior-ceiling surfaces. I hid these fixtures by adding a 1x6 trim detail to the inside top edge of the sidewall beams. Finished with exterior-grade fir plywood and painted to match the main house, the porch soffits are open to the interior. The gable end is also open, eliminating the need for any special soffit or ridge ventilation.

Curved Columns Resemble Tree Trunks

At the gable end, an 8-ft. overhang provides a covered section for sitting on the patio, in addition to shading and keeping wind-blown rain from getting into the porch. This deep gable overhang rests on two curved columns I made in my shop. The columns are double 2x4s cased with primed and painted #2 cedar. I got the curved look by gluing and nailing wedge-shaped blocks of 1x stock to the tops of the 2x4 cores. I ripped both edges of cedar casing so that it was tapered, then glued and nailed the casing to the cores. The bottom of a finished column is a 5-in. square; the top is an 8-in. square. The casing conforms to the posts, giving the columns a curving profile similar to that of a tree trunk.

Spacing the Roof Boards Creates a Skylight

As the roof structure was nearing completion, I realized that although the porch would be comfortably shady during the heat of summer, it might end up a bit too dark. So before installing the last several feet of 2x6 roof boards, I began experimenting with ideas for a ridge skylight. Rather than cut an opening or a series of openings into the roof deck, I decided to take a hint from the lines of the 2x6 roof boards. In the top 2 ft. on both sides of the peak, I separated the boards after ripping off the tongues on a table saw. Starting with a ¼-in. gap, I widened each joint between boards to 2 in. at the peak (see photo p. 174).

Over these spaced roof boards I installed ⅜-in.-thick double-wall Lexan sheets, which might be described as plastic, see-through cardboard. Lexan is obtainable from commercial-plastics distributors. First I folded each 4x8 sheet in half along its length, which was similar to folding a sheet of cardboard. Then I set the sheets in beads of silicone on top of the roof framing, leaving a ½-in. gap between the sheets for expansion. I then filled the gap with silicone.

I fastened the Lexan to the roof structure with battens made of ⅛-in. by 2-in. flat aluminum bar stock. I bent the aluminum to fit over the ridge, set the aluminum in silicone over the Lexan, and drilled pilot holes through the aluminum and the Lexan into the roof framing. Then I screwed the bar stock to the frame, locking the Lexan skylight in place. This simple but effective skylight allows enough light into the porch to brighten it without overheating it. The resulting light from the spaced 2x6 roof boards is similar to sunlight filtering through tree branches.

Finished with exterior-grade fir plywood and painted to match the main house, the porch soffits are open to the interior.

He didn't run out of roofing boards. Because the roof has such deep overhangs, a ridge skylight was needed to brighten the porch. This unconventional skylight is a series of progressively wider gaps between roofing boards, all covered with a Lexan ridge cap. The trusses and 2x6 roof boards are finished with semitransparent stain similar to the gray of the floor and the stonework.

Trimmed Wall Posts Hold Site-Built Screens

For the trim on this project, I used a mix of cypress, red cedar, and white cedar that I primed on all sides with an oil-based primer. Cedar and cypress have natural resistances against rot and decay. All of the trim was finished with an acrylic latex paint to match the existing house finish.

With the framing trimmed out, I was left with a pattern of 5-ft.-square screen openings over smaller 5-ft. by 2-ft. screen openings. I made all the screens using aluminum frame stock and aluminum screen. I made most of the screens on the ground and popped them into their openings. Then I held them in place with 1x2 stock set with finishing nails. If a screen becomes damaged, the 1x2 is removed and the screen comes right out.

In the upper section of the gable end, I had to fit screens into the triangular openings of a truss. I couldn't use manufactured square corner clips to join triangular frames, and the screen, which has a square cross mesh, wrinkled when I stretched it diagonally. I ended up screwing the frames in their openings and stretching the screen in place.

The southwest side of the porch is relatively exposed to both weather and neighbors. Here, I had a roll-down canvas storm shade installed to keep out windblown rain as well as the harsh southeast light in summer. The shade also is a privacy screen, blocking the porch's view from the neighbors' yard.

Tinted Marine Varnish Protects Strip Flooring

For the floor, I chose ¾-in. by 3½-in. tongue-and-groove (T&G) fir to achieve a smooth, attractive, and easy-maintenance surface. The added advantage of T&G boards for the floor surface is that they create an insect barrier, eliminating the need for screening around the skirt or under the joists. There's no plywood. The flooring runs directly over the joists.

The flooring's color, however, was overly reddish next to the bluish gray of the granite patio. So I decided to mix a tinted but still transparent high-gloss finish. After some research and experimentation, I chose a marine-grade spar varnish as the primary sealing element because it's waterproof and UV-resistant. I mixed the varnish with a small amount of bluish gray stain to help deaden the reddish tone of the fir. I applied four coats to the floor, resulting in a high-gloss, semitransparent finish with an attractive grayish tint. The only drawback I discovered to mixing the stain into the varnish was that it seemed to slow the curing process. The varnish remained fragile, even though it was dry to the touch, for about two weeks.

With open-air porches it is common practice to pitch the floor approximately ⅛ in. per ft. away from the main building to allow standing water to drain. With this porch, however, I chose not to pitch the floor so that I could keep all of my trim lines and screen openings level and parallel. I thought that the flooring's marine finish, the large roof overhangs with 4-in. aluminum gutters, and the roll-down canvas storm screen would provide adequate protection to the interior from splashing water. This combination has proved itself effective.

As a second line of defense against water collecting on the porch floor, I installed ⅜-in. brass bushings on 5-ft. centers around the perimeter of the floor to serve as inconspicuous floor drains. These bushings function reasonably well; in the future, however, I probably will use a ¾-in. bushing to allow any intruding water to drain more easily.

Alex L. Varga a one time laborer, carpenter, builder, and now architect, is based in Los Angeles, California.

A roll-down canvas storm shade installed on the southwest side of the porch keeps out windblown rain as well as the harsh southeast light in summer.

Skylight Kitchen

DARK WOOD CABINETS, BROWN AND ORANGE VINYL floor, avocado appliances, yellow tile counters, and a four-tube fluorescent light box. Those were the highlights of the kitchen in the house that we bought 12 years ago in California's San Joaquin Valley. We loved the house, set on a 1-acre lot in a semirural area with mature landscaping and a feeling of spaciousness. But that kitchen had to go.

We made a pass at it a few years later with a new paint job, some new appliances, and a small island with a granite top. But this interim kitchen was just that. We still needed space for two cooks, more daylight, cabinet doors that were not dirt collectors, and work surfaces that were beautiful and functional.

Assembling a Team the New-Fashioned Way

Having decided to go for a full-blown remodel, we contacted an interior designer we found through www.improvenet.com, a service we learned about in *Fine Homebuilding*. We had only one response to our Improvenet query: Marlene Chargin. But you need only one if it's the right one. Marlene referred us to several of her clients and their kitchen projects. We were impressed with the work and with the other members of her team: kitchen designer DeAnn Martin and builder Mark Fletcher. Reassured by the references, we signed up this trio to direct our remodel.

Plenty of light. A flared skylight well lets daylight reach just about every corner of this kitchen. Photo taken at A on floor plan.

Skylight softener. Two layers of fiberglass separated by a grid cut down on thermal gain through the skylight and soften the light falling on the kitchen surfaces.

Clean and simple. Thermofoil coatings are plastic laminates that stretch around corners and edges to minimize dirt-catching seams. This piece, shown with its protective film peeled back, will be a drawer front.

Marlene and DeAnn listened to what we wanted, steered us clear of some pitfalls, and made us feel that the proposed kitchen would really be "our" kitchen. The builder also communicated with us and with the designers well, keeping us informed of daily progress and anticipating design and construction problems before they required reworking.

Easy Decisions, Hard Decisions

We easily settled on a layout that involved bumping out the existing kitchen into a covered patio area (see floor plans p. 181). This plan added about 112 sq. ft. to give us a total of 320 sq. ft. for the new kitchen— not huge, but comfortable for two cooks.

A 2-ft. by 6-ft. skylight was also an easy decision. It is centered over the island in a splayed opening that lets daylight into the entire room (see photo p. 177). Made by Distinctive Skylights (see Sources), the skylight includes a double layer of translucent fiberglass panels separated by a grid (see photo top left). The grid pattern imparts an unexpected design element, the fiberglass panels provide some thermal insulation, and the light is softened by their translucence.

Granite countertops (Blue Pearl from Norway) and tile for the floor were natural choices. Reusing the refrigerator, dishwasher, and dual ovens was also an easy decision. We had a much harder time choosing the cabinetry style and finish.

Our neighborhood is surrounded on three sides by orchards, so dust is a chronic problem in our house. Raised-panel doors with crevices and corners are difficult to keep clean here. Besides, we prefer a sleek, contemporary look. Marlene suggested rigid Thermofoil (RTF), a heat-formed laminate that wraps around the corners of cabinet parts. We went to the local home-improvement showrooms to look them over and saw samples that looked cheap and tacky. Then Marlene located some RTF samples by Prémoule of Ontario, Canada (see Sources). One of their finishes, Silver Ash, was light and silvery, with a subtle wood-grain pattern we loved.

We went about securing a bid for our job from an out-of-town company that specializes in Prémoule cabinets. Their price came in at 75% more than our local cabinet shop wanted for wood cabinets. So we went to plan B. Our local shop farmed out the fabrication of all doors, drawer fronts, and exposed cabinet panels directly to Prémoule. This persistence paid off when the cabinets came in at the original bid price for wood.

The result is a light, clean look. The Thermofoil material is seamless, even at corners, making for easy cleanup (see photo bottom facing page).

Glass in Unexpected Places

The challenge in the cooking area was the treatment of the backsplash. Marlene suggested glass of some sort. Plate glass, either glossy or sandblasted, was ruled out because

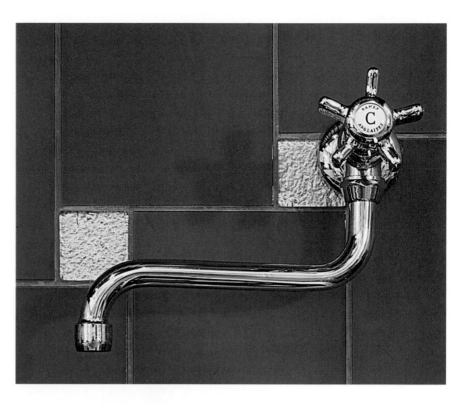

A stepped pattern with small silver glass-tile accents complemented the shape and finish of the exhaust hood.

A splash of brilliant blue. Etched-glass tiles spread out from the center of the cooktop, amplifying the highlights that flicker out of the granite counters. Photos taken at B on floor plan.

SOURCES

www.improvenet.com

ISLAND LIGHTING
MonoRail, by Tech Lighting
(847) 410-4400
techlighting.com

SKYLIGHT
Distinctive Skylights
(800) 430-0076

THERMOFOIL CABINET COMPONENTS
Prémoule
(866) 652-1422

WALL TILE
Azure by Lake Garda
Available through Ann Sacks
(800) 278-8453
annsacks.com

Now there's room for a breakfast table. Bumping the kitchen toward the backyard allowed the island to move away from the dining room. In the foreground, a custom-fit chopping block goes over the sink opening. Photo taken at C on floor plan.

A KITCHEN WITH A SKY-LIGHTED ISLAND

Annexing a little more than 100 sq. ft. of the backyard patio expanded this kitchen into a room large enough to include a breakfast table and a door to the patio. And placing a 2-ft. by 6-ft. skylight over this kitchen's island ensured plenty of daylighted workspace for two cooks.

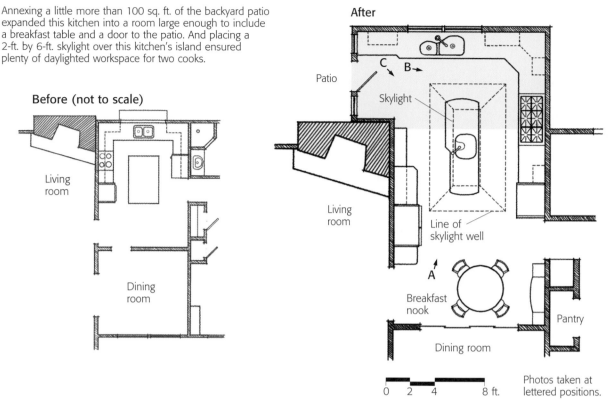

Before (not to scale)

Living room

Dining room

After

Patio

C

B

Skylight

Living room

Line of skylight well

A

Breakfast nook

Pantry

Dining room

0   2   4   8 ft.

Photos taken at lettered positions.

---

it is too hard to keep clean. We finally settled on etched (more of a satin finish) glass tiles. They passed the grease-cleanup test, which was to apply no-stick cooking spray to the tile, then clean with a glass cleaner to see if there was any residue or streaking. With sandblasted glass, some of the oil remained; glossy glass had streaks; etched-glass tile cleaned just fine.

A stepped pattern with small silver glass-tile accents complemented the shape and finish of the exhaust hood (see photo bottom p. 179). At the centerline of the stove, the bookmatched tile patterns come together at a strip of granite that matches the counters.

The island also presented a problem that wasn't resolved until late in the project. We had originally planned to install two appliances in the island: a warming oven and a Kohler combination prep sink/pasta cooker.

The shipment of this component was delayed several times until after the island granite top was temporarily installed. At that point, we realized that the Kohler sink/cooker would eat up more of the island surface than we wanted. We were able to cancel the Kohler order and substituted a smaller prep sink. A pot-filler faucet replaced the pasta cooker.

We got a cutting board at the local kitchen-supply shop and shaped it to fit the island's sink opening (see photo facing page). That extended the island's counter space so that we can work without getting in each other's way.

George Burman and Patricia Looney-Burman cook together at their home in Madera, California.

We settled on a layout that involved bumping out the existing kitchen into a covered patio area. This plan added about 112 sq. ft.

Credits

The articles compiled in this book appeared in the following issues of *Fine Homebuilding*:

p. iii: Photo by Charles Bickford, courtesy *Fine Homebuilding,* © The Taunton Press, Inc.

p. v: Photo by Duo Dickinson

Table of Contents: p. vi (top to bottom)—photos by Laura Du Charme; Doug Dalton; Steve Culpepper, courtesy *Fine Homebuilding,* © The Taunton Press, Inc.; Scott Gibson, courtesy *Fine Homebuilding,* © The Taunton Press, Inc.; Scott Gibson, courtesy *Fine Homebuilding,* © The Taunton Press, Inc.; p. 1 (top to bottom)—photos by David Ericson, courtesy *Fine Homebuilding,* © The Taunton Press, Inc.; Tom O'Brien, courtesy *Fine Homebuilding,* © The Taunton Press, Inc., drawing by Gary Williamson, © The Taunton Press, Inc.; Andy Engel, courtesy *Fine Homebuilding,* © The Taunton Press, Inc.; Charles Bickford, courtesy *Fine Homebuilding,* © The Taunton Press, Inc.; © Rich Zeigner

p. 4: A Coastal Remodel Triumphs over Limits by Duo Dickinson, issue 144. All photos by Chris Green, courtesy *Fine Homebuilding,* © The Taunton Press, Inc., except photo p. 5 © Duo Dickinson. Drawings by Ron Carboni, © The Taunton Press, Inc.

p. 12: A Town House Opens Up in Philadelphia by Tony Atkin, issue 115. All photos by Charles Bickford, courtesy *Fine Homebuilding,* © The Taunton Press, Inc. Drawings by Scott Bricher, © The Taunton Press, Inc.

p. 21: A Top Floor with a Low Profile by Laura Du Charme Conboy, issue 141. All photos by Roe A. Osborn, courtesy *Fine Homebuilding,* © The Taunton Press, Inc., except photos on p. 20–21 © Laura Du Charme Conboy, and photo bottom p. 22 © James Brady. Drawings by Vince Babak, © The Taunton Press, Inc.

p. 28: Elevating the Shingle Style by William L. Burgin, issue 131. All photos by Roe A. Osborn, courtesy *Fine Homebuilding,* © The Taunton Press, Inc., except photo on pp. 28–29 © Doug Dalton. Drawings by Scott Bricher, © The Taunton Press, Inc.

p. 38: Making Room by Todd Remington, issue 108. Photo on p. 38 (inset) by Todd Remington; photo pp. 38–39 and p. 43 by Steve Culpepper courtesy *Fine Homebuilding,* © The Taunton Press, Inc.; photo pp. 36–37 by Roe A. Osborn, courtesy *Fine Homebuilding,* © The Taunton Press, Inc. Drawings by Vince Babak, © The Taunton Press, Inc.

p. 44: Uncramping a Cottage by Matthew A. Longo, issue 118. All photos © Ron Forth, except photo top p. 45 © Peter Post, and photo bottom p. 48 by Steve Culpepper, courtesy *Fine Homebuilding,* © The Taunton Press, Inc. Drawings by Christopher Clapp, © The Taunton Press, Inc.

p. 50: Adding on, but Staying Small by Harry N. Pharr, issue 120. All photos by Charles Miller, courtesy *Fine Homebuilding,* © The Taunton Press, Inc. Drawings on p. 53 © by Harry N. Pharr. Drawings on pp. 58–59 by Scott Bricher, © The Taunton Press, Inc.

p. 60: The View Tower by James Estes, issue 125. All photos by Charles Miller, courtesy *Fine Homebuilding,* © The Taunton Press, Inc. Drawings by Scott Bricher, © The Taunton Press, Inc.

p. 66: Adding a Second Story by Tony Simmonds, issue 102. All photos by Charles Miller, courtesy *Fine Homebuilding,* © The Taunton Press, Inc., except photos on p. 66 (inset) and pp. 68–69 © Tony Simmonds. Drawings by Bob La Pointe, © The Taunton Press, Inc.

p. 74: Expanding a Half-Cape into a Full-Blown House by Joseph B. Lanza, issue 121. All photos by Scott Gibson, courtesy *Fine Homebuilding,* © The Taunton Press, Inc. Drawings by Dan Thornton.

p. 84: Remaking an Old Adobe in the Territorial Style by Ken Wolosin, issue 126. All photos by Scott Gibson, courtesy *Fine Homebuilding,* © The Taunton Press, Inc., except photo on p. 85 (inset) © Jean Clinton. Drawings by Scott Bricher, © The Taunton Press, Inc.

p. 90: A Wish for a Kitchen and a Bath by Louis A. DiBerardino, issue 143. All photos by Tom O'Brien, courtesy *Fine Homebuilding,* © The Taunton Press, Inc. Drawing (in photo) on p. 91 by Gary Williamson, © The Taunton Press, Inc. Drawings on p. 92 by Paul Perreault, © The Taunton Press, Inc.

p. 96: A New Approach to the Kitchen by Andrew Peklo III, issue 120. All photos © Zachary Gaulkin. Drawings on p. 98 by Dan Thornton, © The Taunton Press, Inc. Drawings on p. 100 by Ron Carboni, © The Taunton Press, Inc.

p. 102: A New Master Suite by Scott A. Kinzy, issue 124. All photos © Robert Bately. Drawings by Mark Hannon, © The Taunton Press, Inc.

p. 108: Remodeling a Cape for a Shrinking Family by Howard Pruden, issue 148. All photos by David Ericson, courtesy *Fine Homebuilding,* © The Taunton Press, Inc., except photos on pp. 108 (inset), 109 (bottom), 111 (top and bottom), 113 top right © Howard Pruden. Drawings by Ron Carboni, © The Taunton Press, Inc.

p. 116: Building a Grand Veranda by Kevin Wilkes, issue 132. All photos by Roe A. Osborn, courtesy *Fine Homebuilding,* © The Taunton Press, Inc., except photos p. 116 (bottom left and inset) courtesy Jim Black. Drawings by Vince Babak, © The Taunton Press, Inc.

p. 122: Adding a Covered Entry by Christopher Wuerth, issue 105. All photos by Reese Hamilton, courtesy *Fine Homebuilding,* © The Taunton Press, Inc., except photos on p. 126 © Christopher Wuerth. Drawing by Kathy Bray, © The Taunton Press, Inc.

p. 130: A Shoji Bath by Dale Brentrup, issue 143. All photos by Charles Bickford, courtesy *Fine Homebuilding,* © The Taunton Press, Inc. Drawings on p. 132 by Paul Perreault, © The Taunton Press, Inc. Drawing on p. 133 by Bob La Pointe, © The Taunton Press, Inc.

p. 134: Prairie Home Addition by Ken Dahlin, issue 115. All photos by Andy Engel, courtesy *Fine Homebuilding,* © The Taunton Press, Inc., except photo on p. 139 © Mark F. Heffron. Drawing by Vince Babak, © The Taunton Press, Inc.

p. 140: A Greek Revival Home Gets a New Kitchen by Holly B. Cratsley. All photos by Tom O'Brien, courtesy *Fine Homebuilding,* © The Taunton Press, Inc. Drawings by Dan Thornton, © The Taunton Press, Inc.

p. 148: A Victorian Blossoms in the Orchard by John Seibert, issue 147. All photos by Charles Bickford, courtesy *Fine Homebuilding,* © The Taunton Press, Inc. Drawings by Mark Hannon, © The Taunton Press, Inc.

p. 156: Adding a Sunroom with Porch by Didier Ayel, issue 114. All photos by Steve Culpepper, courtesy *Fine Homebuilding,* © The Taunton Press, Inc., except photo on p. 157 (top) © Didier Ayel. Drawing by Dan Thornton, © The Taunton Press, Inc.

p. 162: Kitchen Remodel, Family Style by Ann Finnerty, issue 127. All photos by Charles Miller, courtesy *Fine Homebuilding,* © The Taunton Press, Inc. Drawings by Paul Perreault, © The Taunton Press, Inc.

p. 168: A Screen Porch Dresses up a Ranch by Alex L. Varga, issue 96. All photos © Rich Zeigner, except photo on p. 169 (bottom) © Alex L. Varga. Drawing by Bob La Pointe, © The Taunton Press, Inc.

p. 176: Skylight Kitchen by George Burman and Patricia Looney-Burman, issue 143. All photos by Charles Miller, courtesy *Fine Homebuilding,* © The Taunton Press, Inc., except photo on p. 178 (bottom) by Scott Phillips, courtesy *Fine Homebuilding,* © The Taunton Press, Inc. Drawings by Paul Perreault.

The articles in this book originally appeared in *Fine Homebuilding* magazine. The date of first publication, issue number, and page numbers for each article are given at right.